Delmar's Handbook of Flowers, Foliage, and Creative Design

Delmar's Handbook of
Flowers, Foliage, and Creative Design

Norah T. Hunter

Africa • Australia • Canada • Denmark • Japan • Mexico • New Zealand • Philippines
Puerto Rico • Singapore • Spain • United Kingdom • United States

NOTICE TO THE READER

Publisher does not warrant or guarantee any of the products described herein or perform any independent analysis in connection with any of the product information contained herein. Publisher does not assume, and expressly disclaims, any obligation to obtain and include information other than that provided to it by the manufacturer.

The reader is expressly warned to consider and adopt all safety precautions that might be indicated by the activities herein and to avoid all potential hazards. By following the instructions contained herein, the reader willingly assumes all risks in connection with such instructions.

The Publisher makes no representation or warranties of any kind, including, but not limited to, the warranties of fitness for particular purpose or merchantability, nor are any such representations implied with respect to the material set forth herein, and the publisher takes no responsibility with respect to such material. The publisher shall not be liable for any special, consequential, or exemplary damages resulting, in whole or in part, from the readers' use of, or reliance upon, this material.

Delmar Staff:
Business Unit Director: Susan L. Simpfenderfer
Acquisitions Editor: Zina Lawrence
Developmental Editor: Andrea Edwards Myers
Executive Marketing Manager: Donna J. Lewis
Executive Production Manager: Wendy A. Troeger
Production Editor: Carolyn Miller
Illustrator: Marcia M. Bales
Cover Design: Carolyn Miller
Cover Image: Marcia M. Bales

Printed in the United States of America
2 3 4 5 6 7 8 9 10 XXX 05 04 03 02 01 00

For more information, contact:
Delmar, 3 Columbia Circle, P.O. Box 15015, Albany, New York 12212-0515;
or find us on the World Wide Web at http://www.delmar.com

Library of Congress Cataloging-in-Publication Data

Hunter, Norah T.
Delmar's handbook of flowers, foliage, and creative design / Norah T. Hunter.
p. cm.
Includes bibliographical references (p.).
ISBN 0-8273-8621-4
1. Flower arrangement. 2. Floral decorations.
SB449.H87 1999
745.92—dc21

99-057012

Contents

Preface

Flower arranging is an art form that enhances the quality of life. Learning the basics and techniques of floral arrangement, knowing how to incorporate the fundamental design principles and elements, and realizing that there is always an opportunity for growth and learning are all essential to becoming a talented floral artist.

The *Handbook of Flowers, Foliage, and Creative Design* is a practical guide for anyone who enjoys the art of floral design. This book provides easy access to basic information as well as step-by-step floral arrangement instructions, a full-color flower and foliage identification section, and a glossary containing technical terminology used within the industry.

Section 1 discusses the basic shapes of arrangements and provides numerous easy-to-follow instructions and illustrations. Section 2 presents a basic to floral design—personal flowers, which includes corsages and boutonnieres. Advanced, contemporary designs and techniques are covered in Section 3.

The Appendices contain extensive listings and illustrations of cut flowers and cut foliage, common to the floral industry. Plants are organized alphabetically by botanical name and include family name, name origin, species, common names, seasonal availability, description of the plant, its use in floral design, and specific care and handling techniques, and approximate vase-life. The common names of plants are listed and cross-referenced to the botanical names because many flowers and foliage are better known by their common names.

The intent of this book is to be a source of information that opens the door to learning and creative, imaginative thinking. Sections contain organized and in-depth examples, illustrations, step-by-step construction techniques, and design ideas. The examples demonstrate the how-to's of floral design in a logical approach to beginning and finishing an arrangement. The book is well organized and written directly to the reader in a helpful, friendly fashion. Hopefully, through reading this book, you will discover within you the motivation to become the best you can be by stretching and expanding your skills, imagination, and desire to build upon your talents.

About the Author

Doug Martin Photography

Norah T. Hunter is a faculty member at Brigham Young University in Provo, Utah, where she has taught students the art of floral design for nearly two decades. She holds both masters and bachelors degrees in horticulture. Ms. Hunter is a member of the American Academy of Floriculture. She recently helped author the book, *A Centennial History of the American Florist*, with *Florists' Review* magazine. She has served as a committee member on the Redbook Florist Education Advisory Board, writing curriculum for floral workshops and seminars.

After working in the retail business for more than fifteen years, Ms. Hunter, a freelance designer, established her own floral consulting company called Norah's Wildflowers. She lives with her husband and seven children in Pleasant Grove, Utah.

Section 1 Floral Arrangements

Floral arrangements today are a blending of two diverse styles—the *oriental style*, or *line design*, and the *European style*, or *mass design* (see Figure 1-1). Often this line-mass blending is referred to as *geometric design* and is commonly known as *Western line design*. Some arrangement shapes are dominantly mass and full, while others are more clearly linear with emphasis placed on line and negative space. Either way, floral arrangements, with the exception of some oriental and abstract designs, can be classified into certain geometric shape groups such as triangular, circular, or vertical.

1-1 *Many of today's floral arrangements are a blending of two diverse styles. The European style on the left is generally round, full, and overflowing, while the oriental style on the right is simple and elegant with emphasis on line. From* Design with Flowers *© 1988 Herbert E. Mitchell; photography: deGennaro Associates, Los Angeles, CA.*

1-2 A successful designer begins with a definite design shape in mind.

The silhouette of an arrangement is its basic shape. It will be helpful for you to become more aware of the basic shapes of arrangements, which will allow you to expand your design options.

Generally, you will find it simpler to achieve a successful floral arrangement by starting with a definite design shape in mind rather than having no plan (see Figure 1-2). This will enable you to choose a proper container, as well as the type and quantity of flowers and foliages appropriate for a particular shape and style. Imagining the silhouette of your design before you begin construction will also expedite the process of arrangement, allowing you to become more efficient with supplies as well as time.

1-3 A specific location and viewing angle will help determine whether a floral arrangement needs to be all-sided or one-sided.

Factors Influencing Arrangement Shape

Many factors influence the choice of a design shape. First and foremost is the placement or location for the arrangement. A specific location and viewing angle will help dictate what shape and size your arrangement will take and whether the arrangement needs to be *all-sided* (designed to be seen from all sides) or *one-sided* (designed with a front, to be viewed from one side). If an arrangement is designed for a coffee table, for instance, it is often long, low, and all-sided. A floral design intended for a fireplace mantle can take on a number of different shapes, all of which are generally one-sided (see Figure 1-3).

The table size and shape on which the floral design will be sitting is also an important factor in determining an arrangement's form. A round table is enhanced with a rounded arrangement. Likewise a rectangular dining room table is an ideal setting for an all-sided, horizontal design. A small bedside table against a wall dictates a smaller arrangement, and often one that is one-sided. However, a bouquet made for the center of a spacious hotel lobby needs to be all-sided and larger in scale, often circular in form (see Figure 1-4).

1-4 *The form of a floral arrangement is determined in part by the table size and shape on which it will be placed.*

The level at which the floral arrangement is viewed can also be a factor in determining the shape of an arrangement. For example, a bouquet displayed in a tall container that rests on the floor will certainly have a different shape compared with an arrangement that sits on a ledge at the top of an arched doorway, as shown in Figure 1-5.

The kinds of flowers and foliages you select will also influence your choice of design. Some flowers lend themselves to tall arrangements, while others do not. Your choice of container can also affect the overall design form. A tall vertical container implies a tall vertical design, while a round basket generally suggests a different form, such as a circular floral design.

The occasion and purpose for a bouquet are also factors that often influence the shape a design must take. The formality of an occasion can also be a governing factor for choosing a shape. *Symmetrical design* is often associated with formal settings, as found in churches, funeral homes, audi-

1-5 *The level at which the arrangement will be placed also influences the shape of an arrangement.*

1-6 A formal setting frequently calls for arrangements that are symmetrical in form.

toriums, and hotel lobbies (see Figure 1-6). *Asymmetrical design*, on the other hand, can lend a more casual feeling and is often associated with less formal occasions and settings. Remember that the overall shape of a floral arrangement must be best suited for the occasion.

Basic Shapes of Arrangements

The basic shapes of arrangements illustrated in Figure 1-7 are the most common geometric shapes used in florists' designs. After considering the requirements of a bouquet, choose one of these forms that will best suit the needs for your arrangement and use it as a guideline for setting up the framework for your design. It is important that you visualize a completed floral design before actually arranging so that you will know how to begin construction.

Most floral arrangements can be considered *triangular, circular, vertical*, or *horizontal*, based on their geometric form. However, some arrangement shapes, rather than fitting into one shape category, are really a blending of several forms (see Figure 1-8).

Once the basic outline of the arrangement is determined, keep your chosen shape in mind, especially as you set up the *framework* or basic structure of the design. The framework for a design can be constructed with foliage or flowers, as shown in Figure 1-9. Generally, the first few flowers or stems of foliage that establish the framework are referred to as *skeleton flowers* or *skeleton foliage*, because they create the main and essential outline for forming a specified design shape. These first few stems set the geometric limits and establish outer boundaries for a design, including height, width, and depth.

To increase the feeling of depth in one-sided designs, it is important to add a skeleton flower or foliage stem in the front of the arrangement near the rim of the container that angles downward. In order to visually balance this downward positioning, angle the tallest stem slightly backward, as shown in Figure 1-10.

Flowers and foliages may then extend slightly beyond the framework edges or they may stay within. Establishing a simple framework will give a foundation for construction and ease the process of arrangement.

Triangular Designs

The triangle shape is a popular choice for floral arrangements. This form has three distinct sides and three corners, angles, or tips. Triangular floral arrangements are generally referred to as one-sided designs, normally viewed on just one side. However, some of these designs can be arranged to be viewed from all sides and still maintain their triangular shape. Many variations of height, width, and size exist, all of which are either symmetrical (formal) or asymmetrical (informal).

Symmetrical Designs

The symmetrical triangle design is formal, and when divided in half vertically, each side of the shape is a mirror image to the other. Remember that when creating symmetrical triangle designs, it is important to keep the framework a mirror image. The flowers within the triangle, however, do not need to be repeated exactly on both sides. This would make an already formal and rigid bouquet much too stiff and unnatural. The basic symmetrical designs offering a traditional formal look include the *equilateral triangle, isosceles triangle*, and *cone design*.

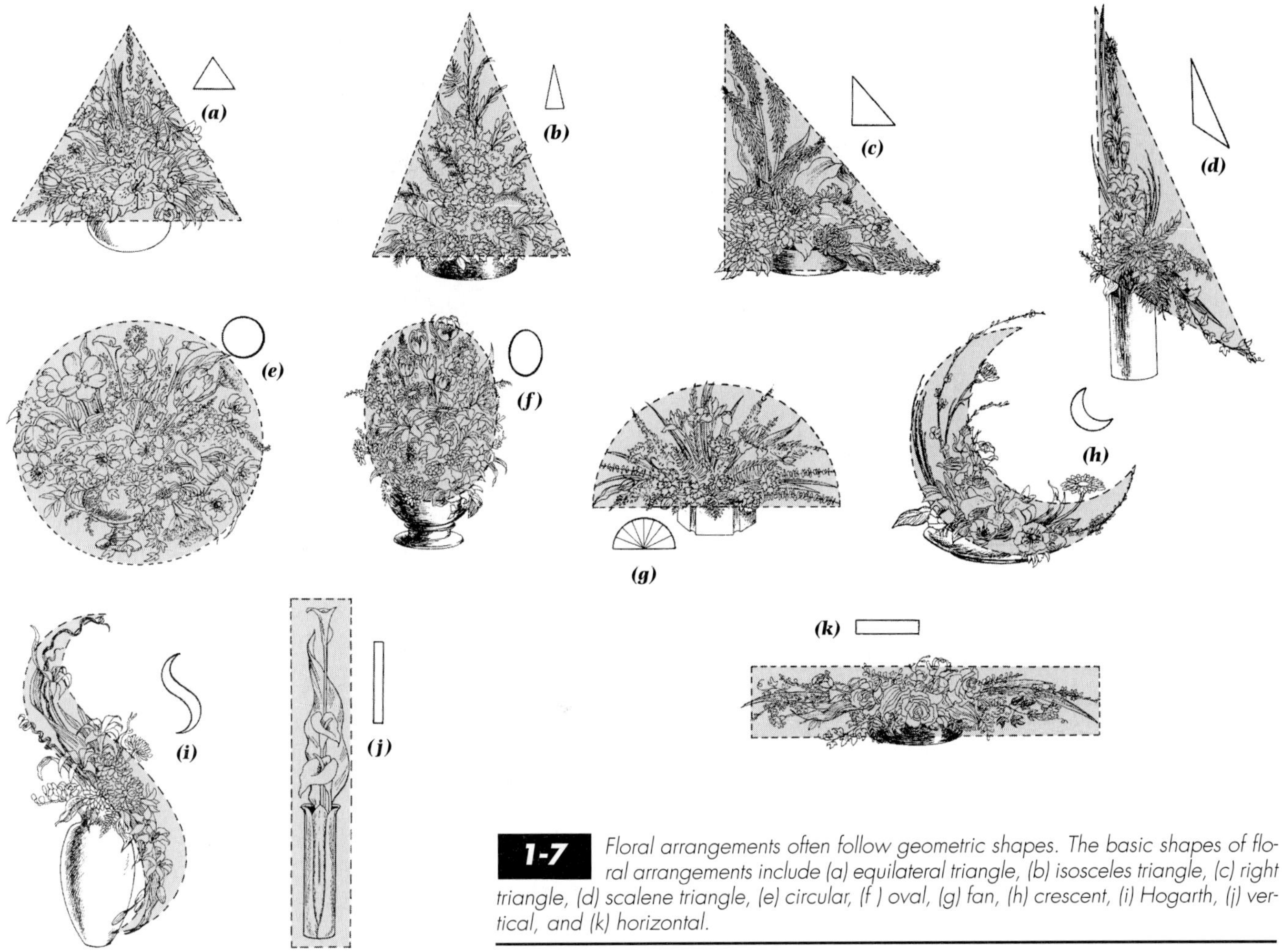

1-7 Floral arrangements often follow geometric shapes. The basic shapes of floral arrangements include (a) equilateral triangle, (b) isosceles triangle, (c) right triangle, (d) scalene triangle, (e) circular, (f) oval, (g) fan, (h) crescent, (i) Hogarth, (j) vertical, and (k) horizontal.

1-8 Some arrangements are a blending of several shapes, rather than falling into a single shape category. The shape of this arrangement of gerbera, iris, delphinium, lisianthus, Queen Anne's lace, waxflower, heather, and statice is based on a circle as well as a triangle.

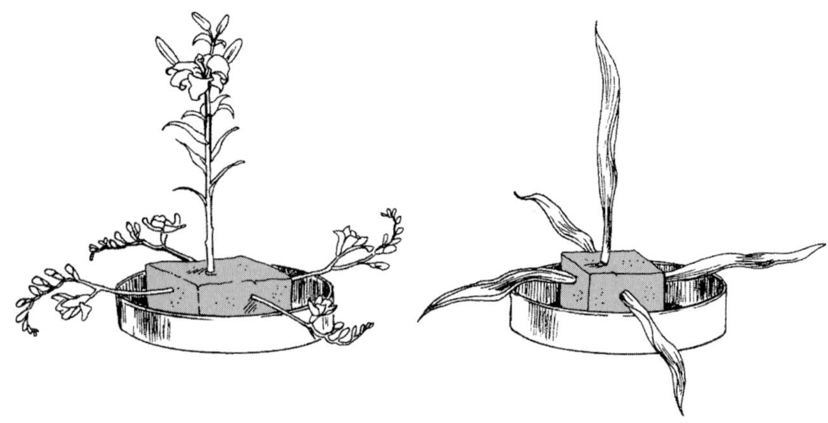

1-9 *The framework, or skeleton, may be constructed with flowers, foliage, or a combination of both.*

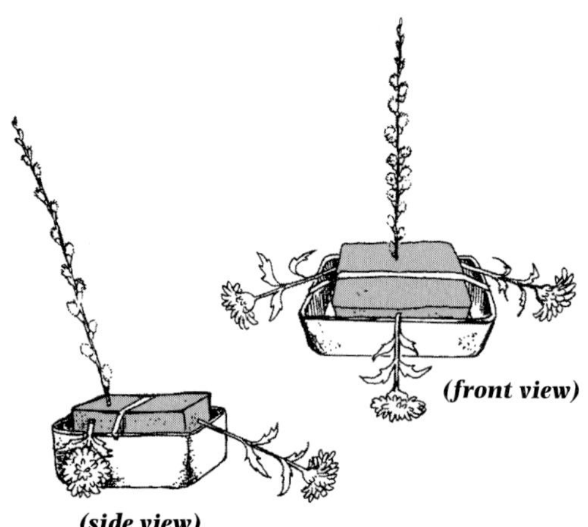

1-10 *As shown on the left, angling the tallest stem slightly backward and placing a skeleton flower or foliage stem in the front of the arrangement angling downward, will add depth to one-sided designs.*

Equilateral Triangle. A triangle that is called equilateral simply means the three sides forming the triangle are all equal in length. These designs are usually one-sided and placed against a wall or in an area where most people will see only one side, such as on a buffet table or at the front of a church.

The equilateral design will generally have three skeleton flowers or foliage stems that create the three points of the triangle. It is important to remember that the three *sides* of the triangle are equal, not the three stems that form the corners of the triangle. The materials used in forming the base line of the triangle need a combined length approximately the length of the stem forming the top center point of the triangle. This dimension will result in the three sides remaining equal to one another (see Figure 1-11).

Isosceles Triangle. An isosceles triangle is also symmetrical, having two sides equal in length with the third or base side unequal (see Figure 1-12). The most common isosceles floral arrangement has two equal sides that come together to form the height of the arrangement.

These one-sided isosceles designs are a popular choice for flower arrangements because they can be easily created from just a few flowers. Whether small or large in scale, these designs generally appear highly stylized.

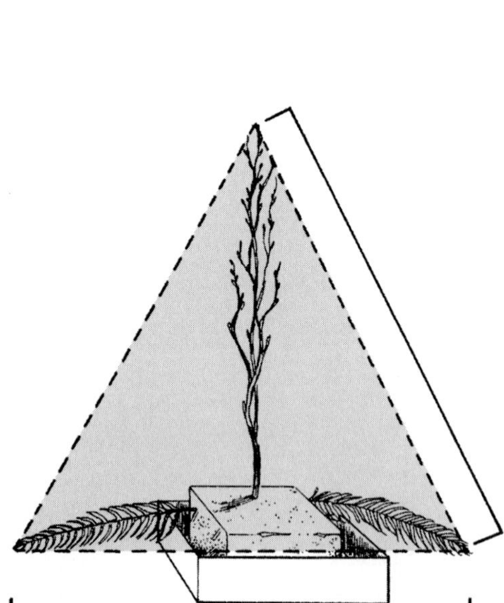

1-11 *Remember, for equilateral triangle designs, the three sides of the triangle are equal (not the three stems forming the corners of the triangle).*

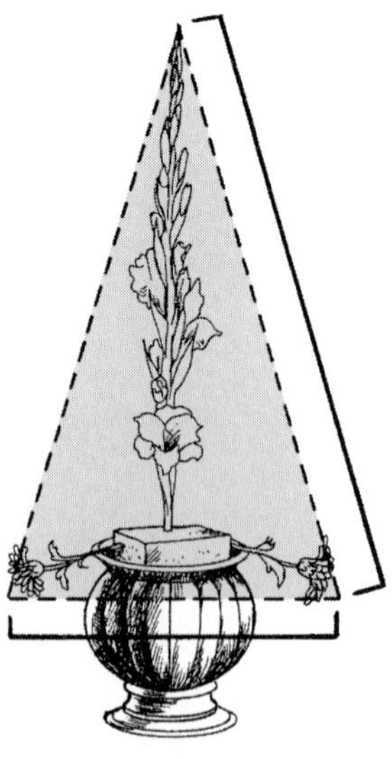

1-12 *An isosceles triangle is symmetrical, with two of the three sides being equal in length. A floral arrangement based on an isosceles triangle is generally a vertical design with the two equal sides forming the height.*

1-13 *For symmetrical triangle designs, secure a piece of soaked floral foam that extends above the lip of the container. To establish the framework of height and width for an equilateral or isosceles triangle, insert three flowers or stems of foliage.*

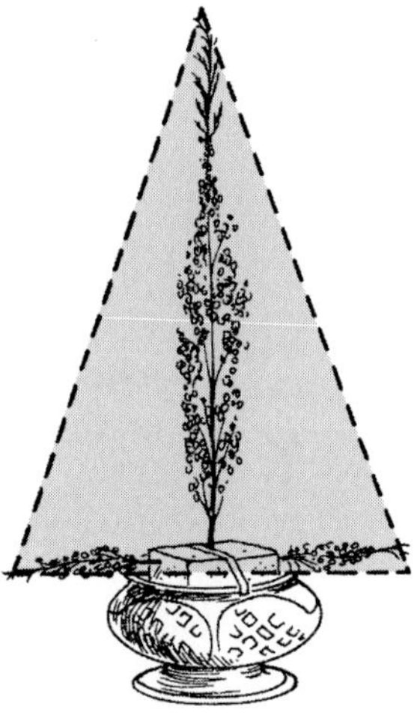

One-Sided Symmetrical Triangle—Steps of Construction

Step 1 Secure a piece of soaked floral foam that extends above the lip of the container (see Figure 1-13).

Step 2 Then add three flowers or stems of foliage to establish the framework of height and width. Since these designs are most often one-sided, to make better use of the floral foam, insert the tallest stem centrally in the back portion of the foam block. Insert the flowers or foliage that create the side triangle points into the sides of the foam, toward the back portion of the foam, so their tips angle forward slightly.

Step 3 Next, add mass flowers to fill the body of the design, as illustrated in Figure 1-14. Complete the triangle shape by adding filler flowers and foliages within the framework you have set (see Figure 1-15).

Remember to sufficiently cover the floral foam not only in front but also in back. Even though these one-sided designs will be viewed mostly from the front, the designer may not know if these arrangements will be placed against a mirror or if some people will see the arrangement only from the back side, as with stage pieces (see Figure 1-16). To achieve a more professional and aesthetic appearance, remember to cover the foam, other mechanical aids, and large unsightly stems with foliage.

1-14 *After establishing the framework of symmetrical triangle designs, add mass flowers and strengthen the desired shape with more foliage and fillers.*

Cone Designs. The cone-shaped design is typical of the Byzantine period style. It actually is a three-dimensional vertical isosceles triangle. These designs require some type of foam base for mass flower insertion. Often pieces of floral foam are gathered together within chicken wire to form these tight designs or a block of floral foam can be trimmed to make a cone. Manufactured floral foam cones are also available. Cone arrangements are formal and rigid and generally appear the same on all sides.

1-15 As the last step to complete your one-sided symmetrical triangle arrangement, add more flowers and foliage within the framework you have set. As shown in this design, all stems—leptospermum, heather, and Scotch broom—seem to radiate out from a central location. The roses face out toward the perimeters, enhancing the feeling of rhythm.

1-16 Although one-sided designs are intended to be viewed from just one side, often, as with stage pieces, they will also be viewed from the back. Always conceal unsightly floral foam, other mechanics, and leggy stems with foliage.

Cone Designs— Steps of Construction

The most time-consuming part of making a cone design is establishing the foam base foundation (see Figure 1-17). The foam must be thoroughly soaked for fresh arrangements.

Step 1 Secure the foam and chicken wire in the container.

Step 2 Add a thin layer of moss with green pins. The moss will easily hide mechanics, and fewer leaves and flowers will be needed to fill in the cone structure.

Step 3 Add fruits, vegetables, berries, or pine cones if desired. Secure with green pins, wire, or wood picks, depending on the individual shape and weight of the items to be added.

Step 4 Insert short-stemmed flowers and foliage pieces through the moss and into the foam to form a colorful cone design, as shown in Figure 1-18.

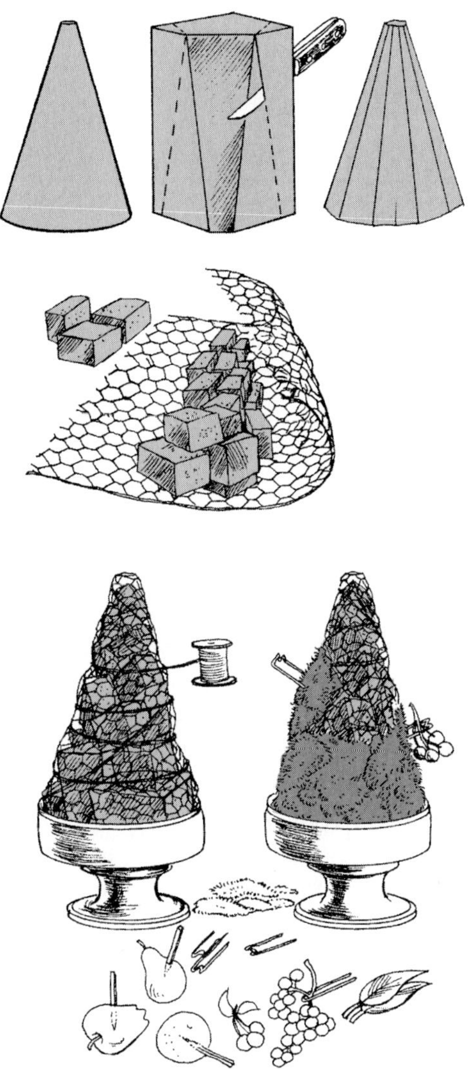

1-17 *For cone designs, a foam cone may be used, a block of foam may be sculpted in the shape of a cone, or several pieces of foam may be gathered together with chicken wire. Chicken wire may also be wrapped around a foam cone, giving the entire arrangement increased support. A thin layer of moss can then be attached with green pins. Start with the biggest elements, such as fruits and vegetables, which can be added to the design by inserting the sharp end of a wooden pick into the fruit and then the other end into the cone. Heavy wires can also be used for added security. Short-stemmed flowers and individual leaves are then inserted into the wet foam.*

1-18 *Once the foundation of a conical design, or an all-sided isosceles triangle, has been established, the design possibilities are endless.*

Asymmetrical Triangle Designs

The asymmetrical triangle design is informal; when it is divided vertically in half, one side is visually heavier than the other. In order to achieve true balance, something must compensate on the side with less visual weight, such as color, texture, form, or the length or angle of a stem.

Asymmetrical designs offer a unique arrangement form. They appear less contrived and more natural. However, more planning is necessary for these designs to achieve proper balance. Asymmetrical (like symmetrical) triangle arrangements are often one-sided but can be made into all-sided designs with initial framework planning.

These design shapes often rely on the use of a focal point for visual success. Asymmetrical triangle designs have the height of the bouquet shifted to one side, and the width of the triangle extending out on the opposite side. The basic asymmetrical designs are *right triangle* and *scalene triangle.*

Right Triangle. A right triangle arrangement is named for the prominent right angle that forms the design. These triangles are generally set in a vertical format with a tall vertical line perpendicular to the base line of the arrangement. The vertical emphasis or height is usually seen on the left

1-19 *Right triangle designs can point to the right or left. The buffet table design set asymmetrically on the table, points to the right.*

1-20 *A right triangle pair will emphasize, frame, and give importance to a painting or other object. The lines at the base of the design may point toward the center or away; either will emphasize the center object.*

side (explaining why these designs are often called *L-shape designs*), but it can just as easily be established on the right side of a design.

Right-triangle designs can point to the right or left, according to the location selected for displaying them (see Figure 1-19). These designs can also be made in pairs and placed on a mantle or table to frame and give importance to a picture or object in between the two arrangements. The right-triangle pair can also be used in an opposite manner (see Figure 1-20). This placement will still give importance to a central object, but at the same time it provides direction and length with the horizontal base line extending outward.

Scalene Triangle. A scalene triangle is one that has unequal sides and angles. A floral design that is made in a scalene form generally has vertical emphasis. This visually impressive design style has an apparent obtuse angle, with height on one side. To create the open angle, the base line extends downward on the other side opposite to the height. Because flower and foliage stems angle downward, it is often necessary to use a *compote* or taller container; if a low container is used, the arrangement should be placed on a mantle or other location where the flowers are allowed to fall freely, preventing blossoms from getting damaged (see Figure 1-21).

1-21 *Scalene triangle arrangements are generally made in vertical containers to allow for downward stems as shown on right. However, they may be made in shallow containers as long as they are located where the downward stems may fall freely as shown on left.*

One-Sided Asymmetrical Triangle—Steps of Construction

Both the right-triangle and scalene-triangle designs have similar steps of construction. Refer to figures 1-22 and 1-23.

Step 1 Extend the floral foam above the container rim to allow for horizontal or downward positioning of stems.

Step 2 Once the floral foam is secure in the container, the framework may be established with flowers or foliage stems.

Step 3 Insert the tallest flower or foliage stem to one side of the design and toward the back of the foam. This placement allows more foam space for the flowers and foliage that will later be placed within the framework.

Step 4 The base of the triangle is formed by *extending* a flower or foliage stem from the opposite side of the design. Insert this stem into the side of the foam block, generally toward the back of the container.

Step 5 To soften the tall vertical line and to help balance the horizontal tip of the triangle, insert a short flower or piece of foliage next to the vertical height of the arrangement.

Step 6 Establish a *focal point* near the rim of the container where tall vertical stems and extended horizontal stems meet. This focal point helps give further balance and stability to the asymmetrical triangular form.

Step 7 Fill in the triangle form with mass flowers and foliage to emphasize the focal area and conceal foam and other mechanics.

1-22 *Right triangle—steps of construction: (a) Secure in the container a piece of floral foam that extends above the container rim. Insert stems to establish the framework for the right triangle, inserting the tallest stem in the back corner of the foam, and opposite to it, insert horizontally a short stem. (b) Next, add foliage near the container rim to conceal portions of the foam and blend the container with the arrangement. (c) Establish a focal area near the rim of the container where the vertical and horizontal lines of the arrangement meet. Add other flowers within the triangle shape. (d) Add fillers and foliage to strengthen the shape, making sure any mechanics of construction are hidden.*

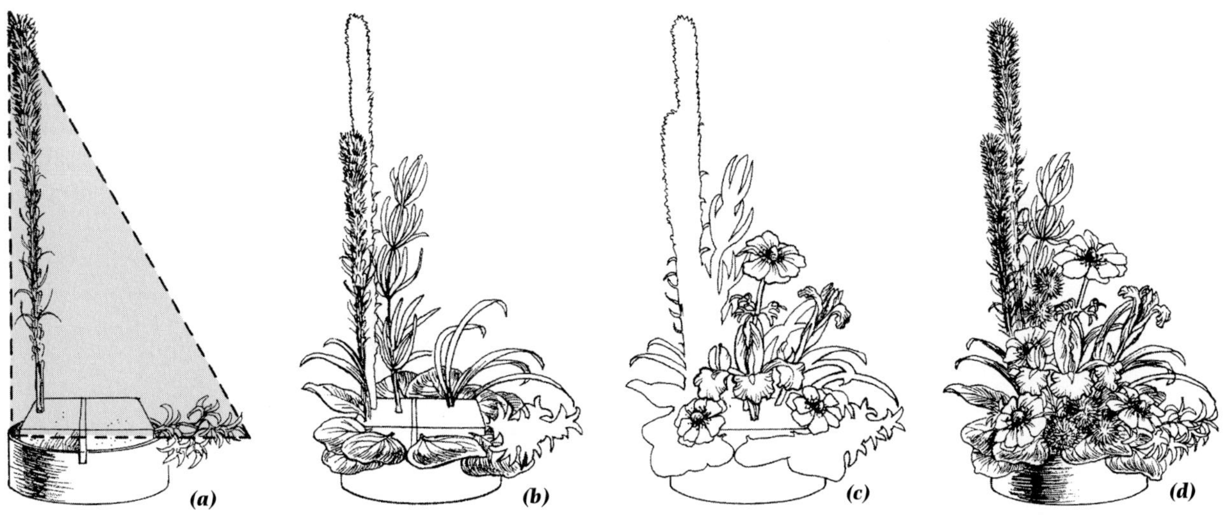

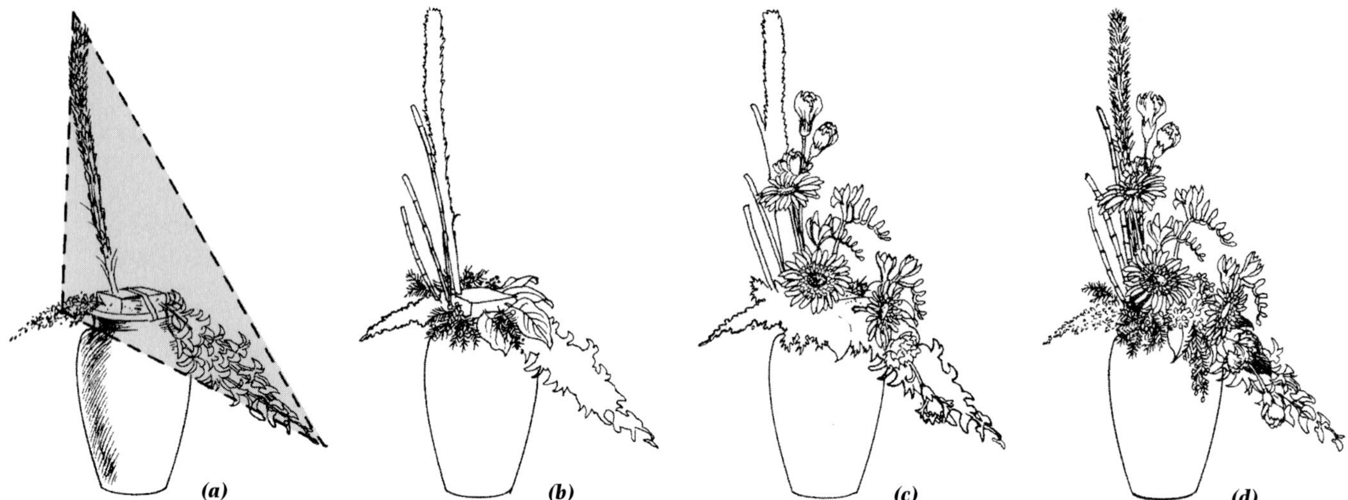

1-23 *Scalene triangle—steps of construction: (a) Generally, a vertical container is required for scalene triangle designs. Secure a piece of foam that extends above the container rim to allow for the downward angling of stems. Establish the framework in the desired scalene triangle shape; the stem establishing the height of the design should be inserted toward the back on one side or the other. The front stem opposite to the height of the design is inserted horizontally into the foam and allowed to angle downward to the side. (b) Add foliage near the rim of the container. Other foliage may be added to strengthen the framework. (c) Establish a focal point and add mass flowers within the framework. (d) Continue adding fillers and foliages to strengthen the shape, making sure any mechanics of construction are hidden.*

Circular Designs

Even though the following arrangements are simply grouped as circular, many diverse styles exist, including symmetrical, asymmetrical, all-sided, and one-sided. Although some circular design styles are more popular than others, it is helpful to know about each and how each may be constructed.

Mass Styles

Designs emphasizing mass with a circular or rounded silhouette are symmetrical and can be made in a variety of sizes and arrangement styles. Most mass designs can be categorized as round, oval, *fan*, or a *topiary ball.*

Round Arrangements. The round arrangement shape appears the same on all sides and from all viewing angles. This design form is also called circular, *round mound*, roundy-moundy, nosegay, and tussie-mussie. Because there is no front or back side to this arrangement, these designs offer more versatility and can be placed in a number of locations (see Figure 1-24).

Arranging flowers and foliages in a circular manner adds an interesting element of repetition that is pleasing and harmonious. However, to avoid the feeling of monotony, use foliage that offers a pleasant contrast to the dominant round flower forms.

If a round or sphere-shaped arrangement is made only one-sided, it usually lacks balance and appears unfinished, as shown in Figure 1-25. When an arrangement must be one-sided, it might be better to select and plan an arrangement shape other than round before beginning construction.

Oval Arrangements. Oval arrangements are symmetrical, yet, unlike round arrangements, they are commonly made in both all-sided and one-sided styles. These designs offer an extension of the round form with the

1-24 *Circular arrangements may be placed in a variety of settings.*

1-25 *As shown from the back, one-sided circular designs lack balance. Even after foliage is added to the back concealing foam and other mechanics, they still will look unbalanced and unfinished.*

circular shape being generally elongated vertically. All of the flowers and foliages are placed as if to radiate out from a central location similar to round designs, the stem lengths being more varied to form the oval.

Fan-Shaped Designs. A fan-shaped arrangement is sometimes called a *radiating design.* These designs are one-sided and similar in shape to a fan or half circle. Often line flowers and foliages are used to set up the framework of the design. Stems should appear to radiate from a central location in the design. These designs are symmetrically balanced with a formal, planned appearance. Voids or negative spaces between the linear framework material give importance to the radiating lines in this mass design.

One-Sided Rounded Designs—Steps of Construction

One-sided oval designs and fan-shaped arrangements, though different in shape, share similar steps of construction. Refer to figures 1-26 and 1-27.

Step 1 Secure foam in an appropriate container. Foam should extend above the lip of the container.

Step 2 Establish the framework. Remember to tilt the tallest stems slightly away from the front of the design. Also, angle the front, lower stems downward for an increased appearance of depth.

Step 3 Add line flowers and foliages to emphasize the radiating appearance of the fan-shaped form. Use a variety of flower and foliage forms to avoid monotony in these circular, mass designs.

Step 4 Continue adding flowers and foliages to form the final shape. Add fillers if desired to soften these concise shapes.

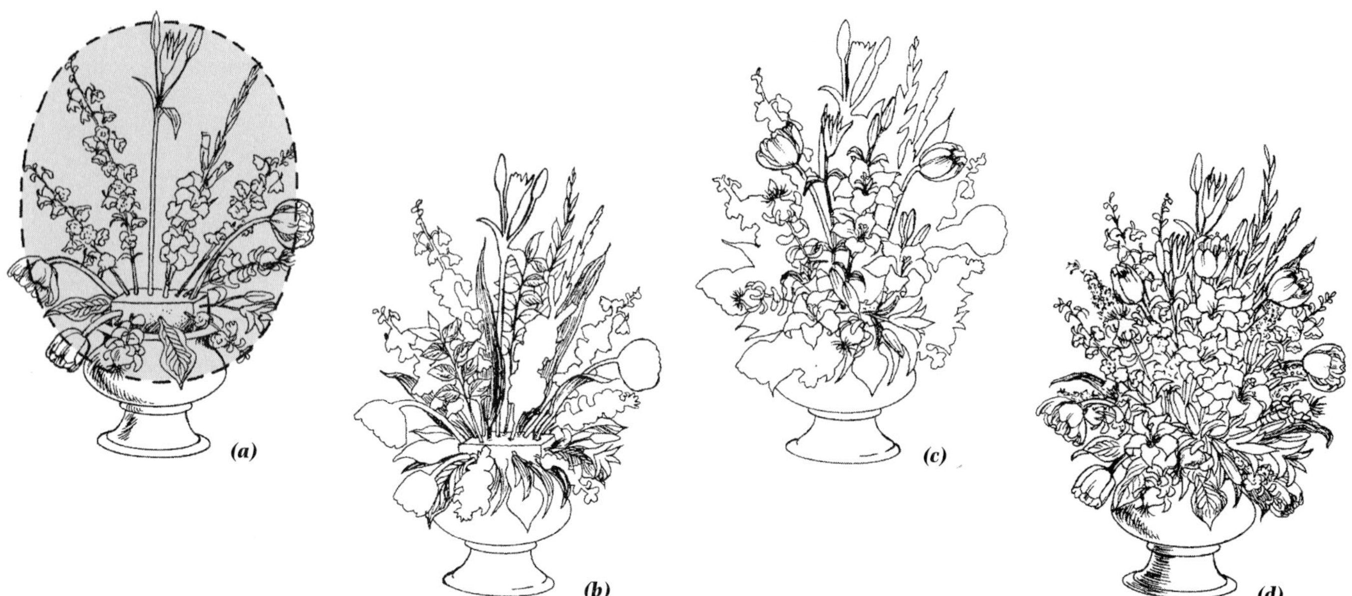

1-26 Oval design—steps of construction: (a) Oval designs are made both one-sided and all-sided. Extend the foam above the lip of the container. Establish the oval framework all around. (b) Add foliage to conceal the mechanics as well as to give a fuller look to the design. (c) Add mass flowers in a radiating fashion within the framework of the design. (d) Continue adding fillers and foliage to the desired thickness and shape. Remember to conceal any mechanics.

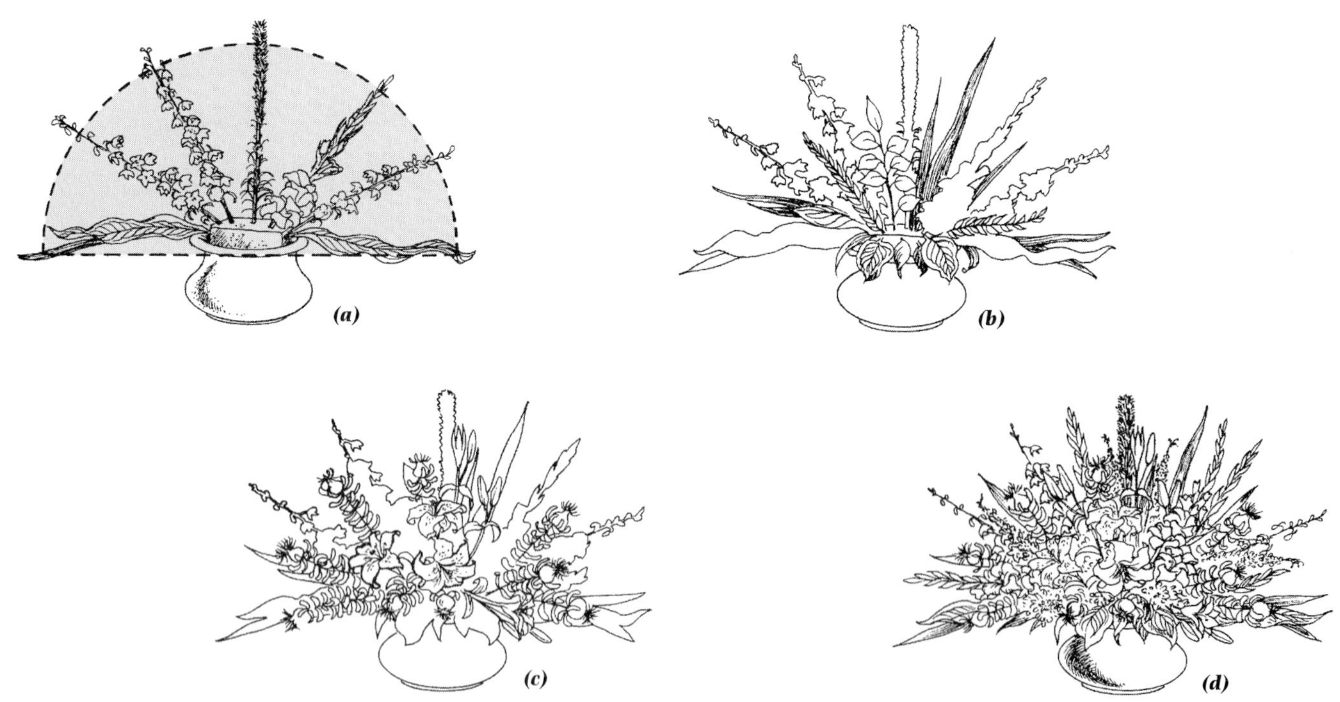

1-27 Fan design—steps of construction: (a) Secure a piece of floral foam that extends above the container rim. Establish the foundation of height and width, as well as the intermediate "spokes" of the fan. Since this is generally a one-sided design, angle the top stems slightly backward for better balance. (b) Add foliage near the container rim and between the framework. (c) Add mass and line flowers in a radiating fashion. (d) Continue adding fillers and foliage, keeping them within the established framework. Conceal any mechanics, especially on the back of this one-sided design.

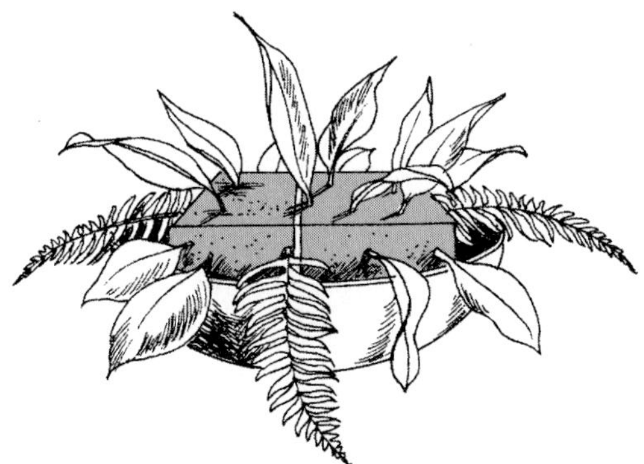

1-28 *Round arrangements are commonly pregreened around the container rim, often referred to as "collaring," before flowers are inserted. This helps to define the shape and cover the floral foam.*

1-29 *When too much foliage is used to pregreen an arrangement, there is little space left for flowers. If pregreening, or collaring, is done before flowers are inserted, the stems must be angled downward in the foam, rather than horizontally. Too many horizontal stems use up foam space and make it difficult to insert the flower stems.*

All-Sided Circular Designs—Steps of Construction

Before beginning a circular arrangement, it is important to consider both the size and mood needed. Choose a container that will offer the appropriate size and proportion relationship with the finished rounded design.

Step 1 Cut floral foam to fit the container, leaving $\frac{1}{2}$ to 1 inch above the rim. Secure foam in the container with an anchor pin or waterproof tape.

Step 2 Often with rounded designs, it is helpful to *pregreen* (also called *collaring*) the arrangement with some foliage before flowers are inserted, as shown in Figure 1-28. This step will help define the shape all around and will cover large portions of the foam and other mechanics of construction. Use only enough foliage to establish the width on all sides, placing stems near the rim of the container.

Step 3 Insert a stem of foliage in the center of the block to help establish height. Be sure to allow open spaces in the foam for flower insertion. If too much foliage is used to pregreen an arrangement, it is often difficult to find any space left in the foam for flowers (see Figure 1-29).

Step 4 Establish the height of the arrangement with a small flower or bud.

Step 5 Next, insert flowers near the rim of the container that extend out horizontally or slightly downward to establish the width on all sides.

Step 6 Insert additional mass flowers to fill out the circular format on all sides. All flowers should radiate from the center of the design. It may be helpful to constantly turn your arrangement or place it on a rotating Lazy Susan so that no one side is neglected, and your design ends up symmetrically circular on all sides (see Figure 1-30).

Step 7 Leave some voids or negative space for visual rest. Voids also allow room for insertion of further foliage to soften the design and to conceal any mechanics still visible.

Step 8 Add filler flowers last to help complete the design shape and style.

Topiary Ball Designs. The topiary ball design is a perfectly round sphere. The shape appears the same from any side or viewing angle. Round topiaries are symmetrical and offer a formal decorating idea in a variety of design styles, sizes, and applications.

1-30 *Round arrangement—steps of construction: (a) Secure a piece of floral foam that extends above the container rim. Pregreen, or collar, the arrangement with foliage all around to establish the rounded form. Establish the framework of height and width on all sides. (b) Add flowers and foliage in a radiating pattern, keeping the outside edges rounded on all sides. (c) Continue adding fillers and foliage in the skeleton, leaving some negative spaces at the edges of the design.*

Topiary Ball— Steps of Construction

The topiary itself is made by inserting flowers and foliage closely together into floral foam.

Step 1 Make the foundation from pieces of foam or sculpture foam and wrap it in chicken wire (see Figure 1-31). Foam spheres and foam holders are manufactured specially for topiary ball construction.

Step 2 Attach the foam ball to a sturdy branch, piece of piping, dowel, or other narrow linear material that is anchored into a heavy pot for use as a centerpiece.

Step 3 After the foundation is secure, cover the ball with a thin layer of moss.

Step 4 Add flowers and foliage stems through the moss and into the foam (see Figure 1-32a).

Step 5 Remember to cover any mechanics, especially in the pot that holds the plaster of Paris or other anchoring material. Mechanics at the base can easily be covered with a thick layer of moss or with an accenting floral arrangement, as shown in Figure 1-32b.

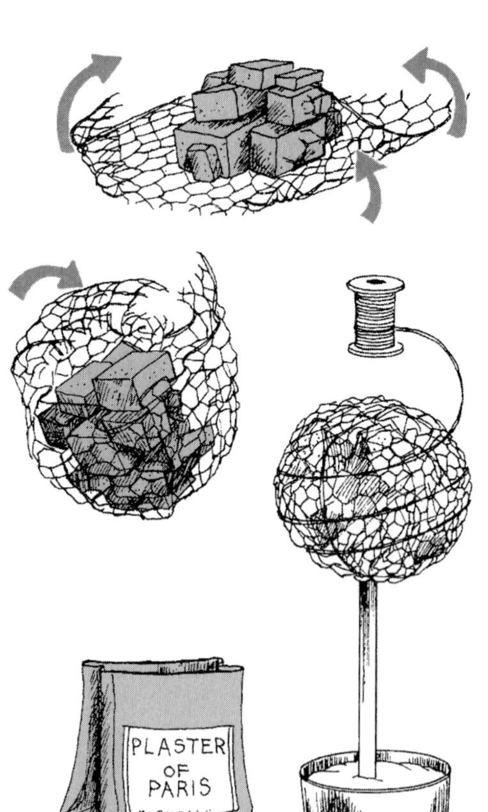

1-31 *The foundation of a topiary ball may be made from pieces of wet foam wrapped in chicken wire. However, foam spheres and foam cage holders, which are manufactured for the purpose of making topiary designs, will save design time. The topiary ball must be elevated by a dowel or branch that is secured in the container. Plaster of Paris is commonly used to anchor the design. Follow the label directions.*

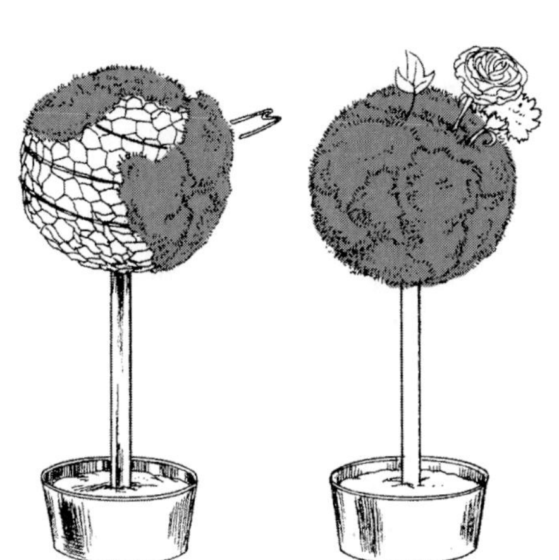

1-32a *After foam is secured to a sturdy branch or dowel, the ball is covered with a thin layer of moss before inserting flowers and foliage.*

1-32b *Cover mechanics at the base of a topiary arrangement with a thick layer of moss or a floral design that harmonizes with and gives accent to the topiary ball.*

Topiary balls have other floral design applications. For instance, they can easily be hung as a room decoration. A small topiary, often called a *floral pomander,* may be carried (see Figure 1-33a). Before flowers are inserted, the ribbon or string that the topiary hangs from must be securely anchored into the ball (see Figure 1-33b).

1-33a *A topiary ball, often called floral pomander, may be hung as a room decoration or may be carried.*

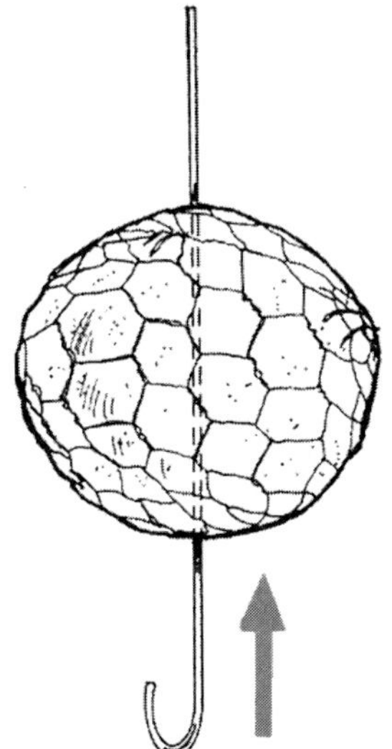

1-33b *To make a floral pomander, insert a heavy wire through the center of the foam ball wrapped in chicken wire. Form a hook on one end and pull the hook back into the foam, catching the chicken wire so that the wire does not cut through the ball. Ribbon or string can be attached through the top of the wire piece as shown.*

Line Styles

Some circular designs are constructed so that the line of the design, rather than the mass, is emphasized. These arrangements are generally highly stylized one-sided designs, but with initial framework consideration they may be made all-sided.

The curved lines of these designs create a unique and sophisticated appearance and display graceful rhythm, resulting in arrangements that are a pleasure to view. Circular line designs that are named for their curvilinear form are the *crescent*, or *C-shape*, and the *Hogarth*, or *S-shape.*

Crescent Design. The crescent floral design is a portion of a circle with one concave edge and one convex edge, similar to the visible shape of the moon just before the first quarter or after the last quarter phases. An arrangement requiring skill and experience, the crescent shape is also referred to as C-shape and can be constructed in a variety of sizes, heights, and widths (see Figure 1-34). Because of its asymmetry, the crescent requires a great deal of negative space with a focal point generally near the container rim for visual balance.

Like other asymmetrical designs, crescent arrangements may be used in pairs accenting and giving importance to a picture or object placed between them (see Figure 1-35).

A crescent design is also beautiful when inverted and symmetrical (or asymmetrical), as illustrated in Figure 1-36. In creating inverted crescent designs, select an upright container and flowing stems to create the downward curves at the sides. All-sided, inverted crescents form a graceful, rhythmic centerpiece, although these designs may also be made as one-sided arrangements.

1-34 *The crescent shape can be constructed in a variety of sizes, heights, widths, and styles.*

1-35 *Like other asymmetrical designs, crescent arrangements may be used in pairs, accenting an object or picture placed between them.*

Crescent Designs—Steps of Construction

Choose a container that will be suitable for the type of C-shape arrangement you will be making. An inverted crescent design generally requires a compote or other tall container, allowing the side stems to hang downward gracefully, whereas an upright crescent most often is made in a low container.

Step 1 Secure a piece of floral foam in the container, allowing the foam to extend above the container rim to allow for stem insertions into the sides.

Step 2 Select curving stems to form the C-shapc. Some stems are naturally curved, while others are not. With hidden mechanics, however, many stems will form the desired tight or loose curve. As shown in Figure 1-37, Scotch broom will easily curve with some gentle persuasion. Hold the stems and blow hot air directly onto the foliage a number of times. These stems will begin to take on the desired curvilinear form. You may also use wires for curving flower stems, but remember to conceal any wiring mechanics you use.

Step 3 Insert the line foliage first to form the curves that display the "C" (see figures 1-38 and 1-39). Because of the length and curve of these skeleton stems, insert them deep into the foam for extra security.

Step 4 Next add some mass foliage that will blend the curvilinear foliage with the massed area of the design. This step will help establish the inner framework as well as conceal portions of the foam.

(*Continued*)

1-36 *An inverted crescent design is graceful and rhythmic.*

1-37 *Many foliages and flowers will provide natural curves to create the crescent line. Others may be made to curve. Scotch broom, for example, can easily be curved by blowing hot air directly onto it while shaping it with the heat of your hands. Flowers and foliages may also be wired to form curves.*

1-38 Inverted crescent design—steps of construction: (a) Select a raised container. Secure floral foam that extends above the container rim. Establish the curvilinear framework. These designs may be made one-sided or all-sided. (b) Add foliage near the rim to conceal mechanics and blend the arrangement with the container. Add other stems of foliage to blend the massed area with the curving linear portions. (c) Add flowers within the established framework. (d) Establish a focal area and continue adding fillers within the framework, concealing mechanics. Do not fill in important negative space.

1-39 Upright crescent design—steps of construction: (a) Secure floral foam, making sure that it extends above the container rim. Establish the framework with curving stems. For one-sided designs, place the stem forming the height of the bouquet toward the back of the foam. (b) Next add foliage near the rim of the container to conceal mechanics and blend the massed area with the curvilinear portions of the design. (c) Add mass flowers within the framework. Establish a focal point at the rim of the container where the height of the bouquet meets with the width. (d) Continue adding fillers and foliage within the established framework. Do not be tempted to fill in important negative spaces.

Step 5 Insert mass and line flowers next. Taper the flower sizes, placing buds and smaller flowers at the outer perimeters of the design. The flower sizes should gradually get bigger or increase in visual weight toward the center of the design. As stems radiate out from a central area, so should the flower heads to enhance the feeling of rhythm.

Step 6 Establish a focal area where the two side portions meet. This area may be achieved simply by massing flowers closer together. A single form flower will also create a focal point as well as give stability to the design.

Step 7 Add mass flowers and mass foliage to form the crescent shape.

Step 8 Complete the arrangement by adding filler flowers and foliage or accessories that will unify the design and complete the desired shape. The crescent form relies on negative space for visual success. Do not be tempted to add more material than is needed (see Figure 1-40).

1-40 *A crescent form relies on negative space for visual success. A design such as this that is too full will lose its crescent shape and look more like a round design with curved lines added.*

Hogarth Curve Designs. Of all the arrangement forms, this design style is the only one that is named for a person rather than a geometric shape. These designs are named for the English artist William Hogarth (1697-1764). Their shape comes from Hogarth's self-portrait titled *Portrait of the Painter and His Pug*, dated 1745, in which the artist drew a serpentine line on a painter's palette with the words, "The Line of Beauty" under it. Many were puzzled with this phrase and curious about its meaning. In response to questioning, Hogarth theorized that all beauty was based on the serpentine S-line, documenting his beliefs later in *The Analysis of Beauty*, published in 1753. A two-dimensional S-line he called "The Line of Beauty," whereas a three-dimensional S-line he called "The Line of Grace." Floral arrangements made in this serpentine line truly display the graceful rhythm for which Hogarth is known.

Often called the S-curve style, these floral arrangements display a sophisticated asymmetrical appearance. Because these arrangements have a downward sweeping curve that extends below the container rim, they must be constructed in a compote, or tall vase. The rhythmic line is easiest to achieve with vines, pliable branches, and naturally curving flower stems. These arrangements offer an elegant option, and they are often the choice of design shape for formal gatherings.

The S-shaped arrangement is generally not as popular as other design shapes because it is more difficult to construct, it requires curving stems, it requires a taller and heavier compote or container, often adding to the expense, and it may easily tip over during delivery. However, the graceful, serpentine line has application in many floral designs and floral decorations (see Figure 1-41). Learning the steps for making a Hogarth design will allow you to choose this design more readily for floral arrangements.

1-41 *Although less popular than other shapes, the Hogarth S-line is lovely and versatile for many decorating purposes such as, weddings.*

Hogarth Curve— Steps of Construction

Step 1 Select an appropriate raised container. Secure a block of floral foam that extends above the rim for horizontal stem insertions. (Refer to Figure 1-42.)

Step 2 Establish the framework with curving linear material. Insert the stem that forms the height into the back portion of the foam on one side. Insert the stem that forms the lower curve of the "S" in the front (opposite to the height of the "S"). Secure these stems deeply into the foam.

Step 3 Similar to the steps of construction for the crescent, next add some mass foliage to soften the curvilinear lines. This step blends lines into the central, massed area of the design and conceals some of the foam and tape.

Step 4 Strengthen the "S" line by adding more linear material.

Step 5 Next, insert mass and line flowers. If a variety of flower sizes is used, place the buds and smaller flowers at the perimeters. Flowers should gradually increase in size so they are largest at the container rim. The visual weight of materials should also become heavier in the central, massed area. All stems should appear to radiate from the center of the design.

Step 6 Create a focal point. For example, mass flowers closer together to create an area of emphasis.

Step 7 Add filler flowers and foliage to finish the design. If desired, add fruits, vegetables, or other accessories to the central area of the arrangement. A cluster of grapes that cascades over the container rim, ombined with other fruits anchored in the foam, offers a classic design typical of the baroque period.

Vertical Design Styles

Standard proportion rules may be easily overlooked when constructing vertical designs. Often height will be emphasized by exaggerating the vertical emphasis, thereby creating a dynamic appearance of strength. A vertical container enhances and strengthens the vertical line in a floral arrangement. These arrangements, with their perpendicular line, are often the choice of design shape when display space is limited or when a strong vertical line needs to be emphasized in a room decoration (see Figure 1-43a).

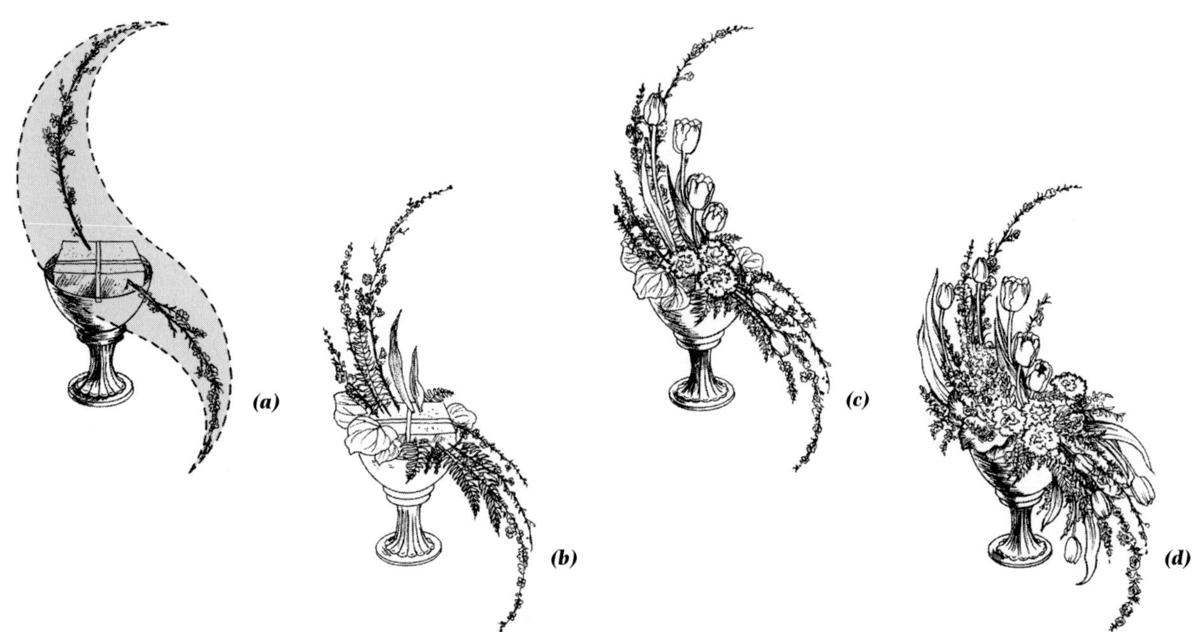

1-42 *Hogarth design—steps of construction: (a) Select a compote or other raised container. Secure floral foam above the container rim to allow for downward positioning of stems. Establish the framework with curving linear stems. Since these designs are generally one-sided, insert the stem forming the height of the bouquet in the back portion of the foam. The stem forming the downward curve may be inserted in the front side portion of the foam. (b) Add foliage all around at the rim of the container and blend the curvilinear lines with the center massed area for a gradual transition. (c) Establish a focal area near the rim of the container where the height and the width seem to intersect. Add flowers and foliages within the established framework. (d) Continue adding fillers and foliages to form the desired shape and style of design. Remember to conceal foam and other mechanics in the back.*

1-43a *Vertical designs are useful when display space is limited or when an emphasis for a strong vertical line is desired.*

Bud Vase Arrangements

The simplest vertical design is a *bud vase* with a single flower or a limited grouping of flowers and foliage. The flowers repeat the shape of the tall slender vase in which they are placed. A bud vase design is generally one-sided, but with initial planning can be made all-sided for use as a simple centerpiece.

Because a bud vase is small in scale compared to other designs, it is usually placed on a small table or in a small room or area. This way the delicate arrangement is not overwhelmed. Placed in a large room or on a large table, a single bud vase is out of scale and tends to disappear. However, bud vase arrangements may be used in groupings for increased visual impact (see Figure 1-43b).

Bud Vase Arrangements—Steps of Construction

Containers used as bud vases are generally tall and narrow. Because the containers are narrow, flowers are easily held in place. Some vases, however, might be too narrow for the number and thickness of the stems intended for a bud vase design. A good "rule of thumb" for selecting a bud vase is this: If you cannot fit your thumb down the neck of the vase, it probably will not be suitable for more than one or two stems of plant material (see Figure 1-44). Forcing flowers into a narrow-necked vase is difficult and frustrating and may crush tender flower stems.

Step 1 Select a suitable vase and fill with tepid water and preservative.

Step 2 Insert the primary flowers to establish height and width. Vary the height and positioning of flowers, as illustrated in Figure 1-45. Keep the number of flowers and the size of the design in scale with the visual and physical weight of the vase.

1-43b *Bud vase arrangements may be used in groupings for greater visual impact.*

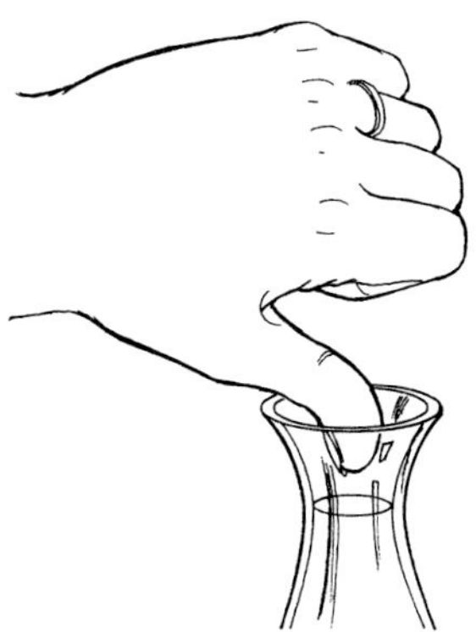

1-44 *When selecting a bud vase, choose one that has a wide enough neck to suit your design needs. If your thumb will not fit down the neck, it might also be a challenge to fit in flower and foliage stems.*

Step 3 Next add some line and mass foliages to enhance the vertical line as well as to blend the arrangement to the container.

Step 4 If desired, add a bow of ribbon, twine, bear grass, or other material to the arrangement at the container rim, to provide accent and visual balance.

Stylized Vertical Designs

Stylized vertical designs are often constructed in vertical containers with the use of floral foam. These designs are generally one-sided but may be arranged to be viewed on all sides.

At times a vertical arrangement may simply have a vertical line of plant material with no emphasis on width. In order to achieve visual interest and balance, these designs often require a focal point either near the rim of the container or at the upper edges (see Figure 1-46).

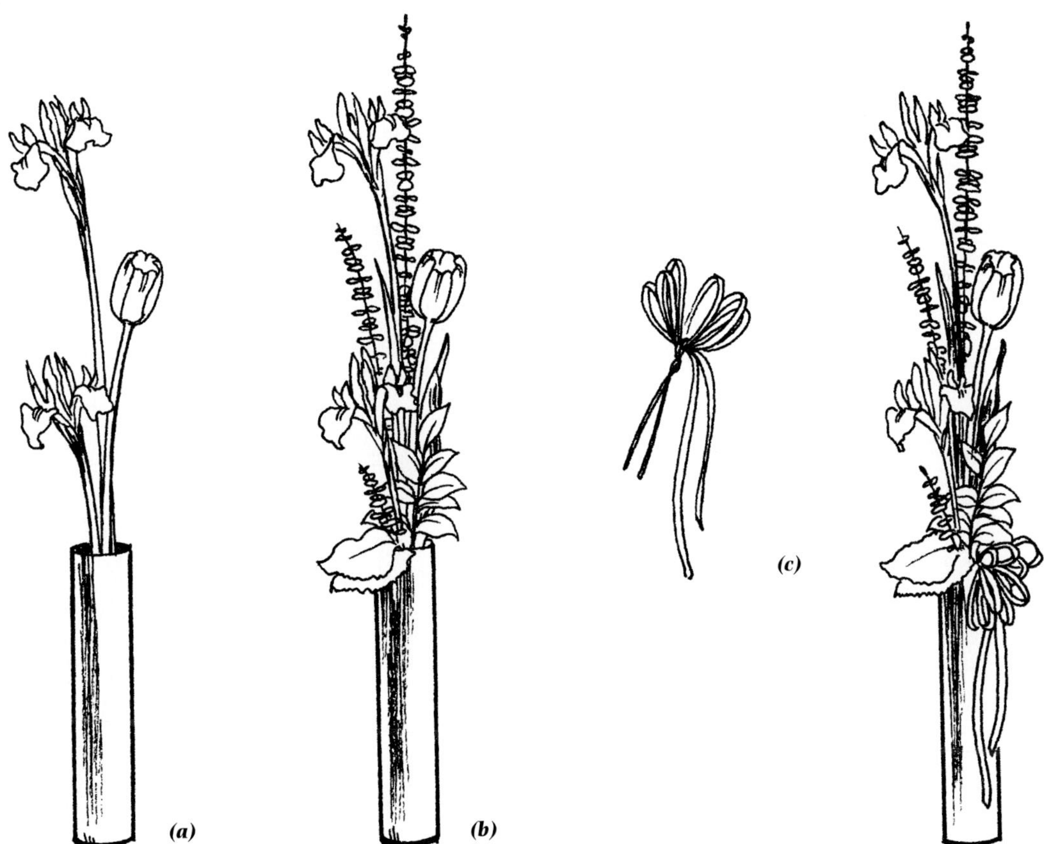

1-45 *Bud vase—steps of construction: (a) Choose a suitable bud vase. Fill the vase with preservative treated water (floral foam generally is not used). Vary the height and position of flowers, keeping the size of the design in scale with the vase. (b) Add foliage near the rim of the vase to help blend arrangement with container. Add line foliage or flowers behind mass flowers to strengthen the vertical line and to help keep flowers securely in place. (c) Filler flowers and foliage may be added to complete and give accent to the design. A bow may be added in front at the container rim or tied around the stems to serve as an accent to the flowers.*

1-46 *Often vertical designs rely on a focal point near the container rim (as shown by the lilies at left), or at the perimeter of the design with an accent near the container rim (as shown by the heliconia at the top and the leaves and curly branch inserted at the rim in the vase on the right). Areas of accent or emphasis are needed to give strictly vertical designs visual interest and balance.*

Stylized Vertical Designs—Steps of Construction

Although these arrangements may be made in low containers, tall containers emphasize the vertical importance of these designs.

Step 1 Select an appropriate container and secure the floral foam if necessary. The style of design, as well as the type of flowers and foliage needed for a particular vertical design, will dictate whether the foam should extend above the container rim or stay below.

Step 2 Establish the height of your arrangement with a line flower or foliage, as shown in Figure 1-47. Since these designs are most often one-sided, insert the first stem toward the back of the block of foam, allowing more foam space for other stem insertions.

Step 3 Add primary flowers first (before mass foliage is added) because foam space is extremely limited in these designs.

Step 4 To enhance the vertical line of the design, add a stem of ivy or other foliage or flower, that extends down in front. This downward line helps visually blend the container with the arrangement.

Step 5 Next, add other flowers and foliages, keeping the line of the design as vertical as possible.

Step 6 Add mass foliage last to accent the focal point and conceal foam and other mechanics.

(a)

(b)

(c)

1-47 *Vertical design—steps of construction: (a) Choose an appropriate container. If floral foam is used, secure a piece that extends above the container rim. Establish the height of the design with linear material. Space will be tight, so position the stem that forms the height toward the back of the foam. (b) Establish a focal point and add foliage to blend the container with the arrangement. (c) Continue adding flowers and foliage, keeping the design prominently vertical.*

Horizontal Design Styles

Horizontal designs, as the name implies, have a strong horizontal line emphasis, one that is parallel with the tabletop or other plane surface. The horizontal line provides a restful, peaceful feeling. These designs are generally symmetrical (but can also be asymmetrical) and can be made all-sided or one-sided. The horizontal line is combined with the triangular or the circular shape, resulting in horizontal designs as shown in Figure 1-48.

Horizontal arrangements are generally made in low and sometimes long containers. Often, the sides of the arrangement will extend past normal width proportion, enhancing the feeling of horizontal line.

This style of design is especially effective when designing flowers for the center of a dining table and is attractive when viewed from any side or angle. Candles are commonly used with this style of arrangement for a formal dinner party. Also, because it is kept low, it will not interfere with conversation across the table (see Figure 1-49).

Horizontal Design Styles— Steps of Construction

Step 1 Select a container that is low for your horizontal design, generally one that is oval or rectangular. A longer container will hold adequate foam and provide a larger area for water.

Step 2 Place the floral foam so that it extends above the container rim so the stems may be inserted into the sides of the foam block.

Step 3 If candles will be used, insert them directly into the foam, or use candle holders. Leave the candle wrappers on the candles to help protect them from scratches during construction. After the arrangement is completed, the wrappers may be removed.

(Continued)

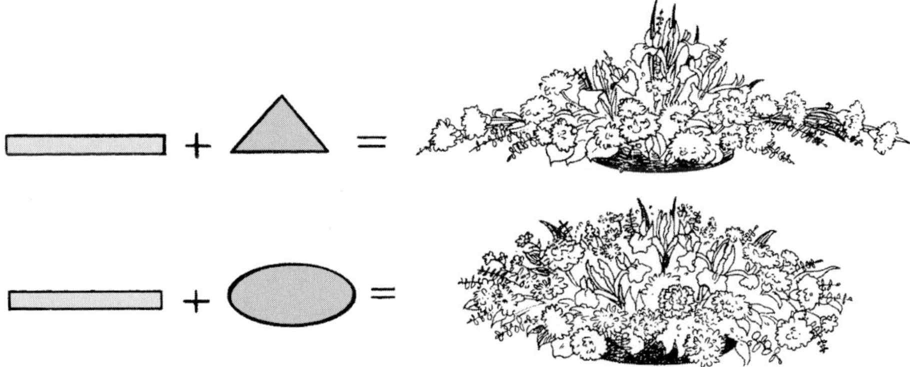

1-48 *Horizontal arrangements are generally based on a horizontal line plus a tri-angular or oval shape.*

1-49 *Table centerpieces should be kept low so they will not interfere with conversation at the table.*

Step 4 Next, establish the framework of the design. Insert flower or foliage stems to establish the length of the arrangement. Insert these stems deeply into the foam, angling them downward, inside the foam if possible, into the water supply.

Step 5 Insert two more stems to establish the narrow width of the framework. Keep these stems shorter, because space is generally limited on the tables where these arrangements are often placed (see Figure 1-50).

Step 6 Place a flower or foliage stem in the center to set the height limit for your bouquet.

Step 7 Add more flowers and foliages near the rim of the container. When viewed from above, these flower heads and foliage tips should extend out far enough to form the pattern of an oval or diamond, as shown in Figure 1-51, to help blend the length of the arrangement with the massed area.

Step 8 Add flowers and foliages to fill in the body of the arrangement and conceal any mechanics (see Figure 1-52). As with S-curve designs, do not be tempted to add too many flowers. Keep the profile of the entire design low.

Step 9 Finally, add filler flowers and foliages if desired to complete the design (see Figure 1-53).

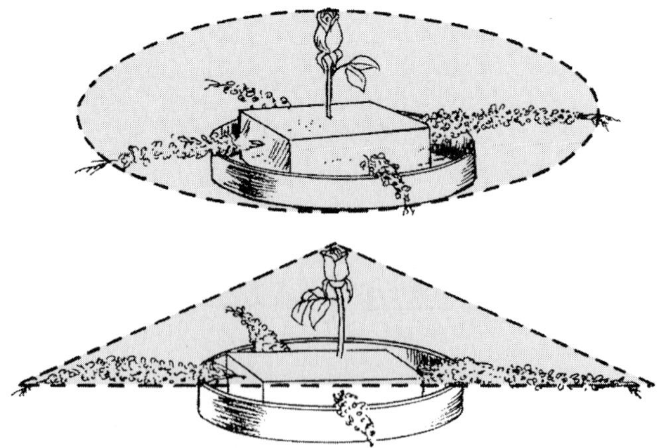

1-50 *Horizontal design—steps of construction: Choose an appropriate low container. Secure a piece of foam that extends above the container rim, allowing for horizontal stems. Establish the framework of height and width.*

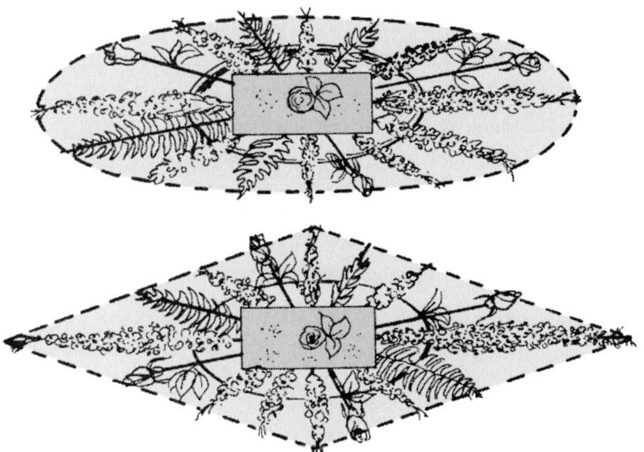

1-51 For symmetrical arrangements, extend flowers and foliage far enough out so when viewed from above, they form an oval or diamond shape, as shown here.

1-52 After setting the initial framework, con-tinue to add flowers and foliage to fill in the body, avoiding the temptation to add too many flowers.

1-53 Horizontal arrangements rely on a low profile. This arrangement of carna-tions, roses, and star of Bethlehem becomes unified as stems of heather, leptospermum, and statice radiate out from a central location.

Most floral arrangements can be classified as triangular, circular, vertical, or horizontal. Learning about these various shapes and the steps of construction for each allows you more design options.

With continued practice, you will become proficient in constructing the basic shapes of arrangements described in this chapter. As you gain more and more experience, your skills and confidence will increase and your desire for further self-expression in the art of floral design can be more fully realized.

SECTION 2 Flowers to Wear

Flowers have long been used for personal adornment and decoration. Modern floral designs meant to be worn or held come from a number of floral traditions. A few examples show clearly that the idea of wearing flowers is not a new one: The ancient Grecian garlands and chaplets, the Polynesian floral leis, the Georgian period formal gown accents, as well as the Victorian Era tussie-mussies as shown in Figure 2-1, all testify to the popularity of using flowers in one's dress.

As illustrated in Figure 2-2, flowers can be arranged in an infinite number of design patterns and worn literally from head to toe. This section provides guidelines of design, wiring and construction techniques, and various design styles.

Guidelines of Design

Many of the principles of design that you follow in making flower arrangements in vases can be directly applied to the art of making corsages and other floral pieces. However, the following principles are additional guidelines for the construction of floral arrangements meant to be worn, often called "body flowers."

Theme and Style

The color and style of the gown, suit, hat, purse, or even hair and hair style to which the flowers will be attached is of utmost importance in determining the color and style of the floral piece. For example, flowers designed to be attached to the décolletage of a sequinned black velvet gown will have a certain theme and style totally different from a floral *chaplet* for a young flower girl.

2-1 *Popular during Victorian times was wear-ing flowers in the hair and on the dress, decolletage, neck, and wrist. Hand held bouquets were fashionable accessory items.*

2-2 *Floral arrangements for wearing may be made in a variety of shapes, sizes, and styles.*

2-3 The parts of a corsage (or other flora piece) must be harmonious to one another as well as for the event for which the design is made.

The occasions or event for which these designs are made is also an important consideration, and knowledge of the environment in which the flowers will be worn will help you select the parts. A formal black tie dinner-dance will most surely dictate a style different from a luncheon honoring the volunteer candy stripers at a hospital. The black-tie event suggests glitz and glamour while a hospital luncheon does not.

The flowers, foliage, ribbon, and other accessory fillers, often called the "parts," used to make a floral piece must be harmonious to one another in color, texture, and style (see Figure 2-3). What kind of flowers you use—whether they are carnations, roses, orchids, or zinnias—is not as important as how well they blend with the other flowers and foliage. Ribbon and other fabric materials added to the design should complement the flowers; the textures, colors, and patterns of ribbon are all elements to consider when adding loops or bows to corsages. Delicate lace ribbon will suggest a different texture and style than that of shimmering metallic gold or silver ribbon. Filler accessories such as hearts, pearls, butterflies, and other tiny novelties, when used, should also be in harmony with the flowers and fit the style.

Proportion and Scale

The sizes and amounts of flowers, foliages, and accessories of the corsage must be in proportion to one another. A corsage with too much ribbon, netting, or other accessories does not allow the flowers and foliage to be fully seen and appreciated.

The size of the completed design should also be in proportion to the person who will be wearing the corsage, boutonniere, or hair piece. This requirement is especially applicable in the case of small children or petite women who can easily be smothered and frustrated by a large floral piece, as shown in Figure 2-4.

Sizes of corsages will vary with trends and styles. Corsages have not always been small and compact. For some occasions, such as homecoming, and in some regions, the philosophy is "the larger, the better." In such

2-4 Proportion and scale are important considerations, especially when designs are made for children. Unlike this corsage, designs should "fit" the wearer.

2-5 Problems may arise if a corsage is the wrong size or shape.

instances, corsages and boutonnieres are designed purposely large and out of proportion to the wearer.

Shape

The basic shape of the floral piece should take into consideration where the design will be worn. For example, the hairstyle will help determine the actual shape of a hair piece. Whether worn on the shoulder, wrist, or neck, the shape of a corsage must be in harmony with its placement. If the shape is wrong for the way the floral piece is intended to be worn, as shown in Figure 2-5, problems will undoubtedly occur. Flowers become a nuisance when they are the wrong shape. Always remember to blend the shape of a floral piece to be an added accessory to the total "look."

Mechanics

Just as you must carefully arrange flowers in a container for proper balance and stability, you must also design corsages, boutonnieres, and other floral pieces securely. The design should be well constructed to retain its original shape, letting nothing fall out of the design (see Figure 2-6). When flower petals shed or parts drop off all together, an embarrassing situation arises not only for the person wearing the floral piece, but for the designer and gift giver as well.

Whether attached with pins, or worn as a wristlet, a barrette, or a clip, a floral piece must be lightweight and easy to wear. Heavy corsages put a strain on clothing and a bulky corsage results in discomfort and self-consciousness (see Figure 2-7). Heavy wrist corsages are undesirable as well, often making the design a burden to wear. A minimum of stems,

2-6 Corsages and other designs to be worn must be secure and well constructed so nothing can fall out and their original shape can be retained.

2-7 *Corsages must be lightweight! Heavy designs cause discomfort and put a strain on clothing.*

wires, and tape will keep the design less weighty. Several guidelines and techniques are listed later in this chapter to help you construct lightweight floral pieces.

Balance

Both visual and physical types of balance are important when making floral pieces. Choosing asymmetrical, symmetrical, or radial balance before beginning construction of a floral piece will help promote mechanical balance throughout construction (see Figure 2-8). Corsages will be more apt to lie flat on the shoulder or wrist when visually balanced.

2-8 *For visual, as well as mechanical balance, choose an asymmetrical, symmetrical, or radial design before constructing a corsage or other floral piece.*

2-9 Flowers that are large, brightly colored, or uniquely shaped will create a focal area, as will accessories in a theme floral piece.

For stability, the heaviest portion of the design should be located at the point where all the stems are physically bound together. This central area should also be the point of attachment to a wristlet, barrette, or wherever the pins hold the floral piece to clothing.

Focal Area

A focal area or center of interest in a corsage or hair piece draws attention and provides a visual, as well as physical area where all lines converge to and from, just as it does in floral arrangements. A focal point can be created in various ways. A larger or more unique flower will easily create a focal point, as will a bright color, dark shade, or any color contrasting with the rest of the design. Sometimes, for special events, accessories used in greater proportion to the flowers will provide an instant focal point (see Figure 2-9). Place the focal point at the center of gravity; never place a focal point at the design edges (for example, flowers at the top and a bow, to serve as a focal point, at the bottom). This treatment generally results in a lopsided design, both visually and physically.

Preparation of Materials

Before construction of corsages, boutonnieres, and other floral pieces begins, the materials going into the design must be prepared. Fresh flowers and foliage must be conditioned, wired, and taped, and occasionally sprayed with paint or an antitranspirant. Accessories such as ribbon loops and tulle fans must be made. Having once done all the initial design work, you will be more efficient during the construction of a floral piece.

Conditioning

Conditioning is a technique that allows flowers and foliages to fully hydrate with water and preservative before using them in designs. Fully turgid or firm flowers and leaves will hold up better after losing their stems to wire and floral tape.

It is especially important to condition blossoms and leaves that are harvested from blooming and green potted plants. Do not wait until the last minute to select various blossoms for corsage work. Give these flowers and leaves time to drink in water and preservative and fully hydrate before using them in corsages and boutonnieres.

Wiring and Taping

Wiring and taping flowers and leaves replaces natural stems; if left on, the stems would be too bulky and heavy. Wire allows more freedom in design, making it easier to maneuver stems and keep flowers in position while being worn. Wire strengthens and, for tiny florets that have been removed from an inflorescence, provides a new stem.

Several wiring techniques are used in corsage work. The wiring method used will depend mostly on the actual shape of the flower head or cluster of flowers. The thickness or gauge of the wire used is determined by the weight of the flower head and where in the design it will be placed. Large or heavy flowers closer to the binding area will require thicker wires. Small, delicate flowers positioned on the perimeters of the design will not require the same thickness of wire.

The most useful wire gauges are medium to fine in thickness. Gauges #24, #26, and #28 are the most common for corsage work. Use the lightest gauge wire possible that will do the job. When making corsages and boutonnieres, trim out excess wire to keep design pieces lightweight. As you prepare your floral piece, be aware that not every addition to a corsage (or other floral piece) must be wired and taped in. Many types of low-temperature glues and floral adhesives are available for adding lightweight accents of ribbon and blossoms.

Roses and Carnations

Roses and carnations have rounded heads and a visible calyx. The most common method of wiring for these and other similarly shaped flowers is often referred to as the *pierce* method (see Figure 2-10).

Step 1 Remove all but about 1/4 to 1/2 inch of stem. Do not remove the green sepals from roses; left on, they provide color and accent, and look more natural.

Step 2 Insert a medium-gauge wire (#24) into the calyx. Bend the two wire ends downward, keeping them parallel to each other. Never twist wires to avoid too much bulk in the stem.

Step 3 Tape the wires by stretching a piece of floral tape around the calyx. Wrap around in one spot until the tape begins to stick to itself.

Step 4 Continue stretching the tape downward onto the wire in a spiral pattern, covering all the wire.

Step 5 After floral taping, the tape can be further tightened by applying pressure with fingertips.

Rose Petals

Individual rose petals may be used to form tiny rose buds for accents and contrast in floral pieces (see Figure 2-11).

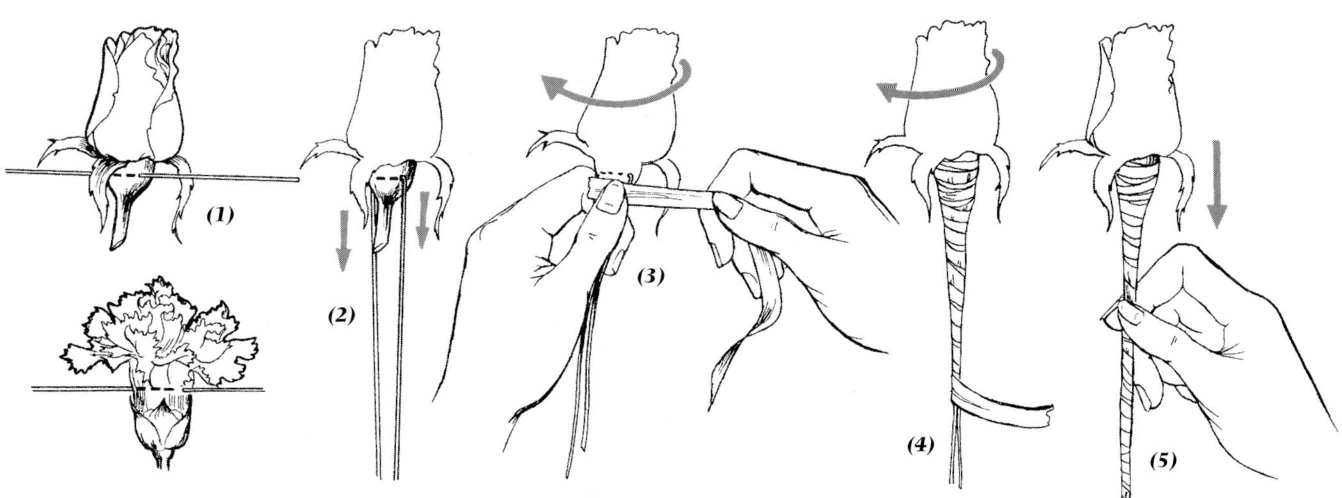

2-10 *The most common method of wiring roses, carnations, and other similarly shaped flowers is the pierce method.*

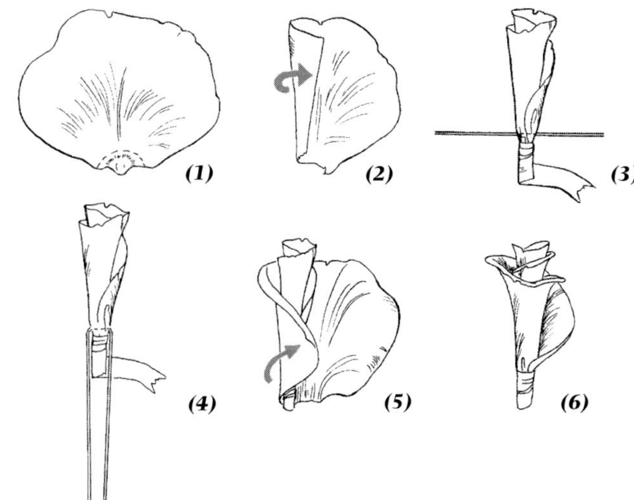

2-11 *A designer may roll individual rose petals to form tiny rose buds for accent or contrast.*

Step 1 To make a rose bud, trim a tiny half circle at the base of the petal in the center.

Step 2 Roll a single petal from side to side.

Step 3 Add some floral tape or low-temperature glue to keep the petal from unrolling. Insert wire above tape.

Step 4 Bend wires down and floral tape.

Step 5 Add a second or third rose petal around the bud to give it more fullness if you wish. Attach these other petals with tape or glue.

Step 6 Use floral tape and tape to form a natural-looking stem.

Feathering Carnations

Petals from carnations may also be used to make smaller flowers. The process of splitting a carnation apart is often called *feathering.* Feathering large carnations takes some time and effort but allows for more versatility in design. Knowledge of this technique is especially beneficial if small carnations are needed for a design and miniature (pixie) carnations are not available. Several flowers (two to six) may be made from a standard-size carnation.

Step 1 To feather a carnation, remove its stem. While pressing firmly at the base of the calyx, roll from side to side until the ovary with its attached center part (the pistil) slips out and can be removed (see Figure 2-12).

Step 2 Next, peel the calyx sections downward, away from the petals. This action will open up the flower head, yet still keep the petals attached to the base.

Step 3 Remove a section of petals to form the new smaller flower.

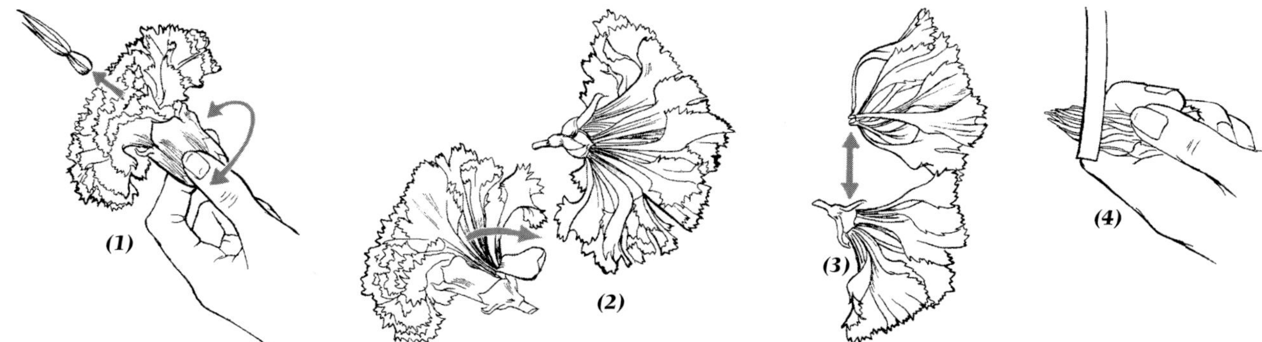

2-12 *To feather a carnation to make smaller florets: Step1. Press and twist (roll) the base of the calyx until the center comes out; Step 2. peel down the calyx sections (like peeling a banana), keeping petals intact; Step 3. divide (or section) pieces of petals; and Step 4. tape base of section together using floral tape.*

Step 4 While holding the petals tightly in one hand, wrap the base of the cluster together with a piece of floral tape. Once the floral tape is tight around the petals (forming a new calyx), the section can be wired.

Step 5 Insert a thin wire just above the floral tape, as shown in Figure 2-13.

Step 6 Bring the wires down together, keeping them parallel to each other.

Step 7 Use floral tape to form a new stem.

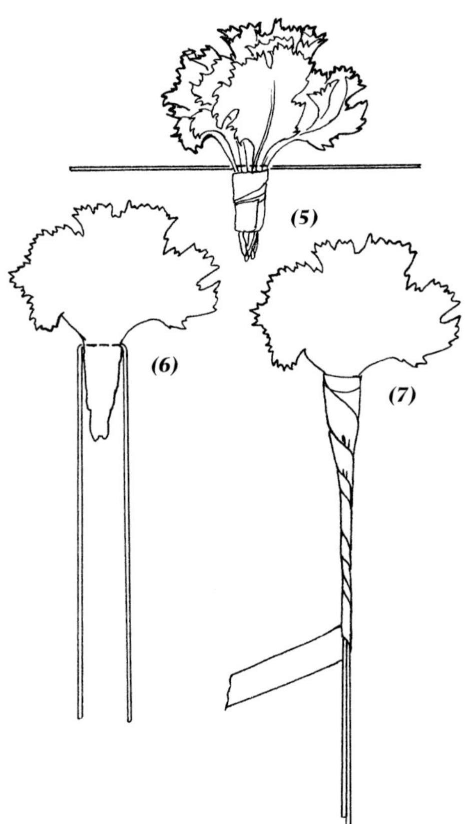

2-13 *As the final steps for feathering a carnation: Step 5. Pierce wire above the floral tape; Step 6. bend the wire down; and Step 7. Tape the wire stem.*

2-14 The hook-wire technique is used for daisies and other flowers with flattened heads and no visible calyx.

Chrysanthemums

Chrysanthemums, asters, daisies, and other flowers with flattened heads lacking a visible calyx are wired for security using *hook-wiring* (see Figure 2-14).

Step 1 Remove all but 1/4 to 1/2 inch of stem.

Step 2 Insert a hooked or U-shaped wire through the top of the flower head, and pull the wire ends down until they cannot be seen from the top.

Step 3 Use floral tape on the wire to form a new stem.

Filler Flowers

Filler flowers and small clusters of tiny mass flowers are often wired by using *clutch* or *wrap-around wiring* (see Figure 2-15).

Step 1 Gather the stems of baby's breath, statice, waxflower, or other tiny fillers into a small cluster.

Step 2 Wrap a lightweight wire tightly around the stems.

Step 3 Bend the wire ends downward, keeping them parallel to one another.

Step 4 Tape the stems and wires together using floral tape.

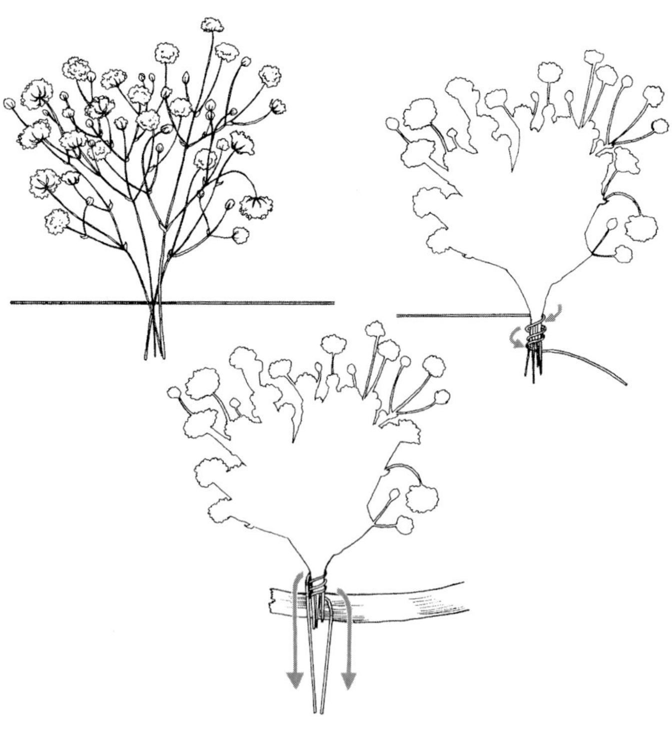

2-15 For filler flowers and small clusters of tiny mass flowers, use the clutch or wrap-around method.

Stephanotis

Stephanotis flowers can be wired in a number of different ways; however, the most efficient method is to use a *stephanotis stem*. These stems are manufactured specifically to provide a stem and keep stephanotis flowers from wilting. Before using stephanotis, it is best to condition them in cool water to firm them up.

Step 1 Moisten the cotton portion of the stem in water for a few minutes.

Step 2 Remove the tiny green sepals and tiny stem from the blossom.

Step 3 Use the wired stem end to push out the tiny ovary through the top of the flower (see Figure 2-16).

Step 4 Insert the cotton stem into the flower. (It can be poked down through the top or can be pushed up through the base of the flower.) Generally, the flower and stem do not need any further wiring or taping and are ready to use in designs.

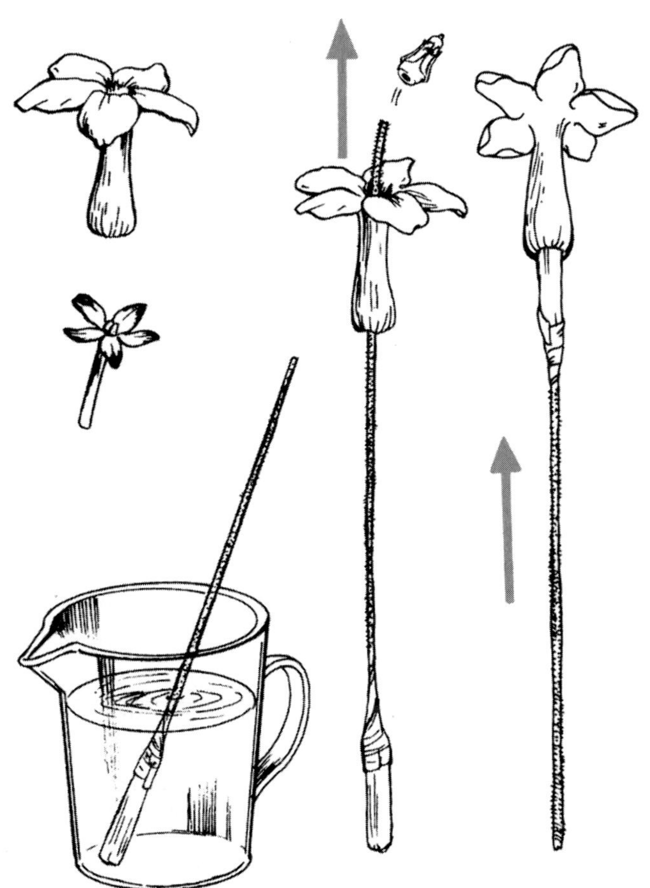

2-16 *Manufactured stephanotis stems may be used to wire delicate stephanotis and keep them fresh while in designs.*

Cymbidium, Cattleya, and Japhet Orchids

Orchids are wired differently than other flowers because of their unique shapes. Orchid stems can be left in a tiny plastic tube for a constant water supply or the stem can be wrapped with wet cotton or tissue (see Figure 2-17). Generally, two wires are needed when wiring cymbidium, cattleya, and japhet orchids.

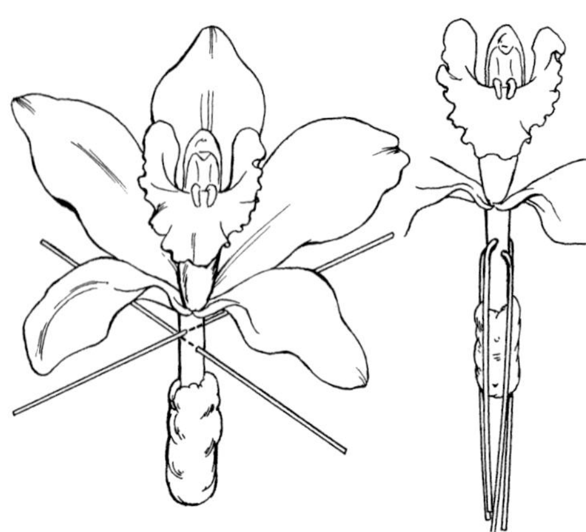

2-17 A wet tissue or cotton held in place by cross-wiring keeps larger orchid flowers fresh.

Step 1 Recut stem and place in corsage water tube, or wrap stem end in a piece of moistened cotton or tissue.

Step 2 Insert a medium wire (gauge #22 or #24) through the orchid stem just below the flower. This wire will provide strength while holding the flower in the proper position.

Step 3 Insert a finer wire (gauge #26) through the stem above the previous wire at a right angle to the first wire. This thinner wire will help keep the moistened cotton, tissue, or water tube in place.

Step 4 Bend all wires downward, keeping them parallel with one another.

Step 5 Tape the source of moisture inside the wires, covering all the mechanics, using floral tape.

Dendrobium and Paphiopedilum (Lady's Slipper) Orchids

There are several ways these orchids may be wired, depending on how they will be used in the design. The stem may be cross-wired with fine wire, as is done for Japhet orchids, or a single medium gauge wire may be inserted vertically into the stem and flower as listed below.

Step 1 Insert a medium wire (#24 or #26) vertically up the short stem.

Step 2 Push the wire out through the throat of the orchid.

Step 3 Curve the wire end into a small hook. Then pull the wire hook gently back into the orchid until it reaches the flower base. The end of the hook will protrude slightly through the back of the orchid.

Step 4 Tape the wires to provide a new, natural-looking stem using floral tape.

Phalaenopsis and Vanda Orchids

Phalaenopsis and vanda orchids are wired differently than other orchids because they are quite fragile and require extra support within the flower head itself (see Figure 2-18).

Step 1 Tape a wire (gauge #24 or #26) with white floral tape.

Step 2 Curve the wire into a hook or U-shape.

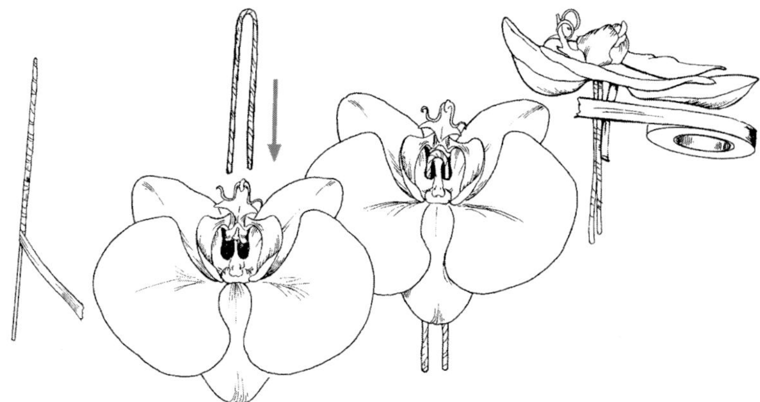

2-18 *Phalaenopsis and vanda orchids require extra support within the flower head itself.*

Step 3 Insert the wire carefully from above the orchid over the middle section and through the visible spaces. The taped wire does not puncture any part of the flower; instead, it provides a source of support for the orchid.

Step 4 Wrap a tiny piece of moistened cotton or tissue around the orchid's stem if you wish. This may be held in place with another fine wire.

Step 5 Using floral tape, tape the stem, wires, and cotton together to provide a longer, more natural-looking stem.

Gardenias

Highly fragrant gardenias are extremely fragile and easily bruised. Browning of petals can be lessened by keeping your hands wet while working with gardenias. Most gardenia flowers are mounted on special *collarettes* with background leaves and packaged in protective boxes that keep them fresh.

To wire a gardenia, it is best to leave the protective collar on the flower. The collar keeps the flower positioned and protects the petals.

Step 1 Insert a wire (#24) through the stem, beneath the collar.

Step 2 Trim the end of the stem and add a moistened piece of cotton or tissue.

Step 3 Insert another wire (#26) through the stem, higher than the first wire and perpendicular to it. This wire will help hold the cotton in place.

Step 4 Bend the wires parallel to the gardenia stem and floral tape.

Step 5 If you wish, trim the leaves on the gardenia collar with scissors.

Lilies and Alstroemeria

Lilies, alstroemeria, and other similar flowers can be wired several different ways. The flower stem thickness and position of the flower (in a corsage or other floral piece) will determine the most efficient wiring technique. Cross-wiring the stem with medium gauge wire works well for large lilies, while the pierce-wiring or wrap-around wiring methods, using fine wire, supports dainty alstroemeria, delphinium, and other small florets.

Step 1 Remove most of the stem except for about 1/2 to 1 inch. Large lilies will benefit by wrapping a small moistened piece of cotton or tissue against the stem end. Remove the anthers (pollen) to prevent staining.
Step 2 Insert a wire (#24 or #26) through the top of the stem. Sometimes a second wire will need to be inserted perpendicular to the first wire.
Step 3 Bend the wire ends downward, keeping them parallel to one another.
Step 4 Tape the stem and wires together, using floral tape, being careful not to damage the flower petals.

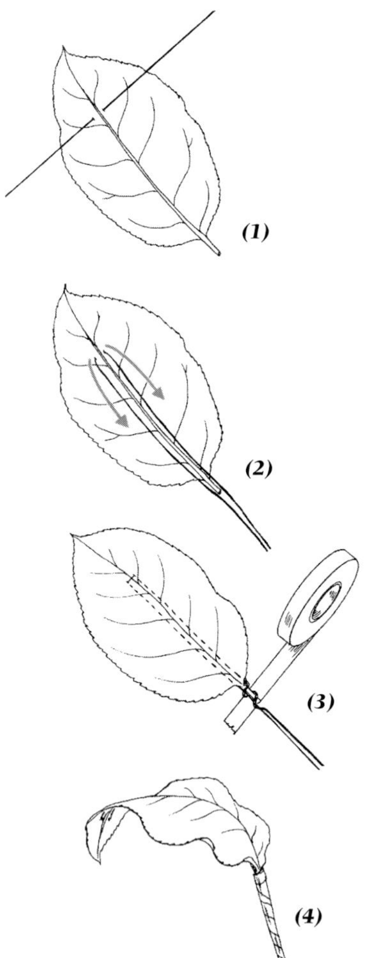

2-19 *The stitch method is used to wire camellia, ivy, salal, and other broad-leaf foliages.*

Broad-Leaf Foliages

Individual leaves of camellia, ivy, salal, and other *broad-leaf foliage* can be wired using *stitch wiring* (see Figure 2-19).

Step 1 Pierce a fine wire (#26) through the back of the leaf near the center vein and make a tiny "stitch." Make this stitch high enough on the leaf to gain control and support the leaf, but not so high that the stitch will be visible in a design.
Step 2 Move the wire until there are two equal ends, and bend them downward.
Step 3 Keep one of the wire ends parallel with the mid-rib of the leaf. Twist the other wire end around both the leaf stem and the straight wire.
Step 4 Tape the leaf stem and wires to form a new, natural-looking stem using floral tape.

Spray Painting

Carnations and a few other flower types can be lightly spray painted with special floral tints. It is important that just the petal edges are painted. This technique is called *tipping* (see Figure 2-20).

Step 1 Wire and tape carnation.
Step 2 Pierce the wired stem through the center of a paper towel (to protect hands from spray paint).
Step 3 Gather the paper towel up around the flower head.

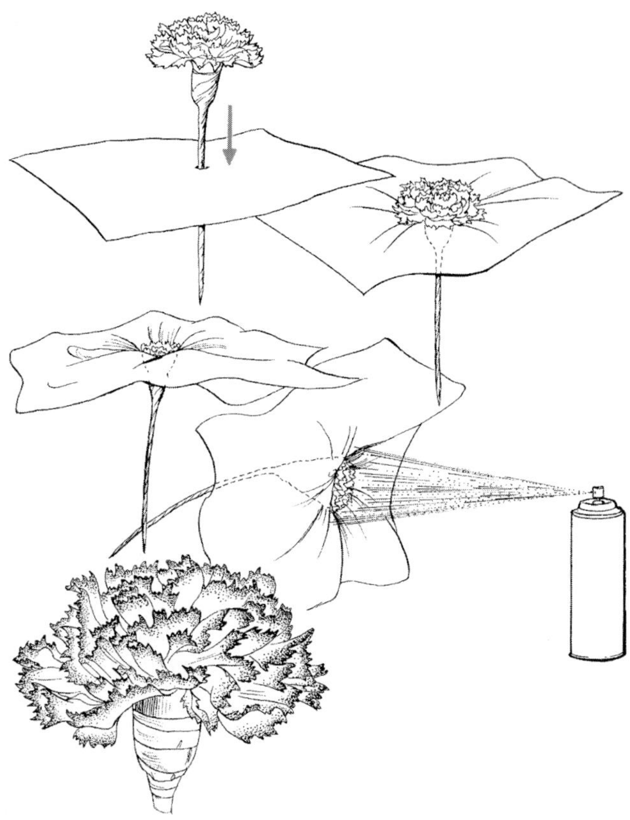

2-20 *When spray painting, or tipping, carnations, be careful to paint only the petal edges.*

Step 4 Squeeze the flower head together with the paper towel tight against the petals. Only the petal tips will be exposed to the paint.

Step 5 Apply an even coat of paint. If the tipped carnation is not dark enough or another paint color will be added, the process can be repeated until the desired effect and color is achieved.

Accessories

Accessories, such as ribbon loops, bows, tulle netting fans, and other novelties, can be added to corsages to enhance a theme and create a unified design. Not all corsages rely on accessories for visual success—some will look better without any extras.

It is important to give thought to the selection of these design extras. Their main purpose is to accent and give importance to the flowers in the design. To keep the entire design lightweight, be sure to choose extras that are not heavy.

Ribbon Loops and Flags

Loops and flags of ribbon are more suited to most corsages than multi-looped bows. A large bow in a corsage can be overpowering and unfitting to the flowers and the entire design in which it is placed. Ribbon loops and flags accent the flowers and add color and texture throughout the entire corsage, helping to unify the whole design. Loops can be made in a number of different ways. The following steps explain how to make a simple single ribbon loop. (see Figure 2-21.)

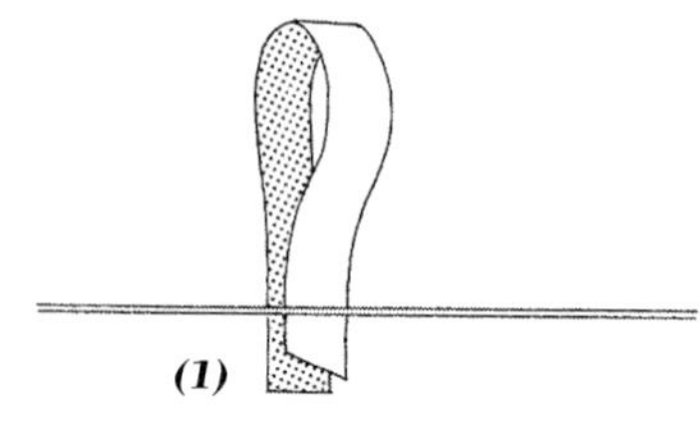

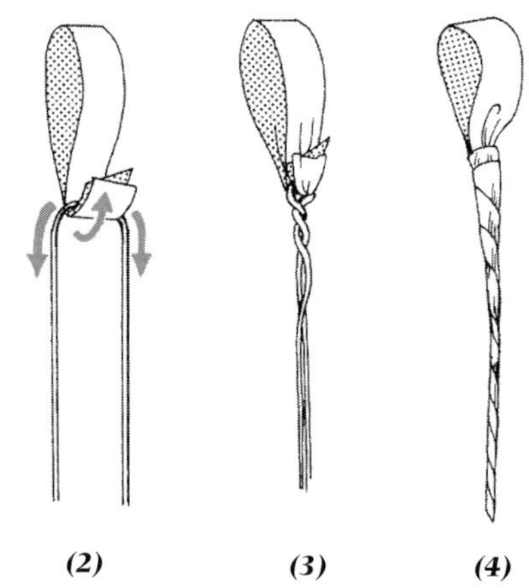

2-21 *Ribbon loops can easily be made with ribbon pieces, wire, and floral tape.*

Step 1 Cut a piece of ribbon 3 to 4 inches long.

Step 2 Fold the ribbon in half with the right side out.

Step 3 Place a wire (#26, #28, or #30) across and on top of both ribbon ends. Fold the ribbon ends up toward the top of the loop.

Step 4 Bend the wire ends downward and twist them together.

Step 5 Secure the base of the ribbon and attach onto the wire, hiding any mechanics by using floral tape.

As shown in Figure 2-22, double and triple loops can be made, as well as loops with tails, called *flags.* A variety of ribbons can add shimmering, glitzy, velvety, or lacy textures and patterns throughout a corsage.

Net and Lace Fans

Net and lace may be added to designs to provide a background and also to create fullness in corsages without adding weight. Net, also known as *netting* and *tulle,* is sold on small bolts, usually 6 inches wide. Netting is available in a wide variety of colors, patterns, textures, and styles. When using netting in corsages, choose materials that are not scratchy and stiff. These can be annoying and sometimes irritating to the skin when placed in shoulder or wrist corsages.

Net and lace can be cut into sections, forming fans, butterflies, or tufts (see Figure 2-23). Several methods can be used to make net background for corsages.

2-22 *Double and triple loops, as well as loops with tails or flags, serve to add color and texture and strengthen unity in a design.*

Step 1 Cut a long piece of netting (12 to 18 inches) from the bolt.

Step 2 Fold the piece of netting in half (from side to side). Fold in half again so the piece is folded into fourths.

Step 3 Cut the net, while still folded, diagonally every $2\frac{1}{2}$ to 3 inches. Trim the ends diagonally as well.

Step 4 Open up an individual piece of net and lay flat.

Step 5 Starting at a straight edge, fold up in $\frac{1}{4}$- to $\frac{1}{2}$-inch pleats (like an accordian).

Step 6 Place a thin wire (#28 or #30) across the top and center of the piece of net (see Figure 2-24).

Step 7 Fold the wire ends down together, while holding the two net sides close together.

Step 8 Twist the wires together. The net will form the shape of a butterfly or fan. These sections may be taped to hide wires and mechanics using floral tape.

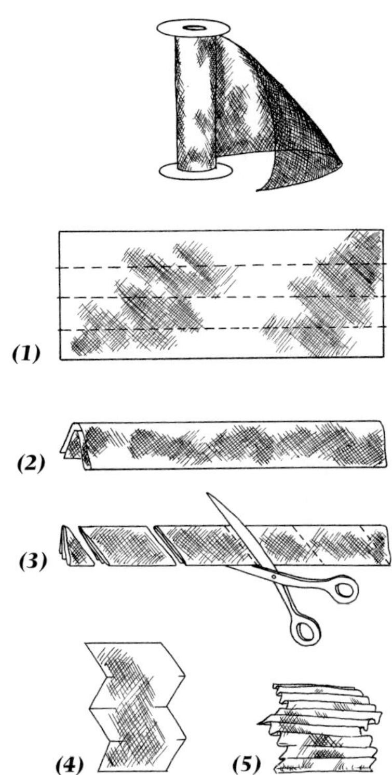

2-23 To make a simple tulle fan-tuft for a corsage: Step 1. Cut a long piece of netting (12 to 18 inches long) from the bolt; Step 2. fold the piece of tulle in half, then in half again; Step 3. cut diagonally into pieces 3 to 4 inches wide; and Step 4. lay the piece flat. Step 5. gather up with ¼- to ½-inch pleats.

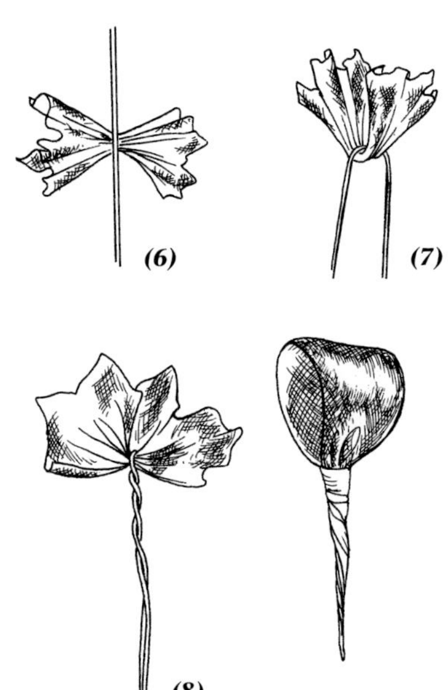

2-24 As the final steps for making tulle fan-tufts: Step 6. Lay fine wire across all layers of tulle; Step 7. fold tulle sides up, while bending wire ends down; and Step 8. twist the wire several times for security. As shown on the lower right, tulle fans can be made in many other ways as well. This tulle tuft is made by folding a piece of tulle and then wiring and taping, resulting in a softer, fluffier tuft of netting.

For a smoother appearance, pieces of netting may also be folded to form a tuft without cut edges. Net pieces are placed to form a background to help visually as well as physically support flowers in a design (see Figure 2-25).

Novelties

Many novelties are manufactured for use in corsages (see Figure 2-26). They range in style from cute, youthful bees and butterflies, to more elegant-looking pearls, rhinestones, and other faux jewels. Lightweight holiday and seasonal novelties are available for use in corsages as well. Novelties generally have a wire attached for convenience in design. However, do not assume that everything must be wired and taped. Low-temperature glues and liquid or spray adhesives may be used to secure some novelties into designs.

Artificial Leaves

Artificial leaves, sometimes called *glamour leaves,* can be added to corsages and boutonnieres in addition to fresh foliage. Artificial leaves are available in a wide range of colors, sizes, textures, and cluster groupings. These leaves add color and texture accents throughout a design and generally add a touch of elegance.

2-25 Tulle works well as a background accessory in corsages.

2-26 *Corsage accessories range from butterflies and bees to pearls and faux jewels as well as holiday and seasonal novelties.*

Boutonnieres

Many events such as weddings, proms, banquets, and other formal events are special occasions for men to wear flowers. However, the occasion need not be formal. Special holidays such as Father's Day, Valentine's Day, or Boss's Day, or sentimental occasions such as an anniversary or birthday are also times when flowers are worn by men. A single flower or small cluster of flowers is a classy addition to a suit coat for any day of the year.

A floral piece worn by a man is called a *boutonniere* and is generally worn on the lapel of a formal jacket or less formal suit coat. Flowers are most often worn on the left lapel near the buttonhole (hence the name "boutonniere"). Smaller pins with black heads (boutonniere pins) or the larger pins with pearl heads (corsage pins) may be used to attach the boutonniere to the lapel. Pins may be inserted from the back side of the lapel if desired.

Whether the boutonniere is a single rose, carnation, a dendrobium orchid floret, or a cluster of stephanotis, statice, or a sprig of holly, the success and security of any design will depend on proper construction techniques. A single flower with some filler and foliage is a popular boutonniere choice (see Figure 2-27).

Single-Flower Boutonniere

Carnations and roses are the most requested flowers for single-flower boutonnieres. There are, however, many other flowers that work well in boutonnieres, such as alstroemeria and delphinium blossoms and dendrobium orchids.

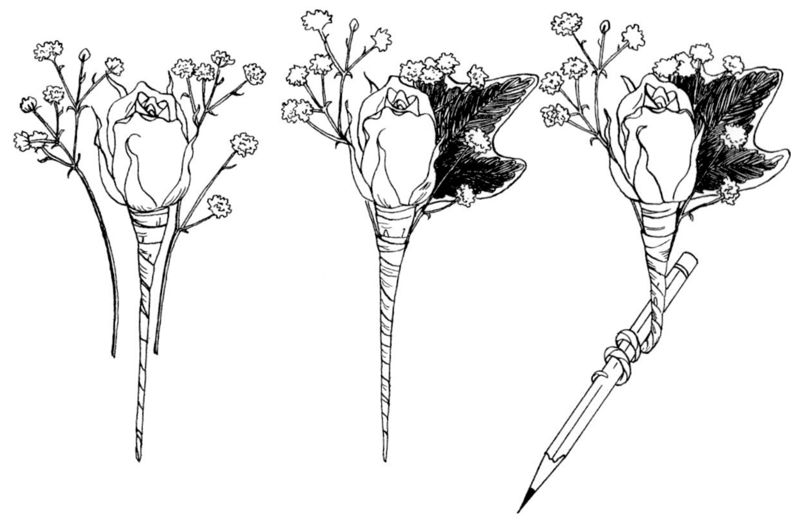

2-27 *After the main flower for a boutonniere is wired and taped, filler flowers and foliage may be added. The stem may be curled easily with the help of a pen or pencil.*

Step 1 Wire and tape a rose (or other flower) using floral tape.

Step 2 Add a sprig or two of baby's breath (or other filler flower, such as statice, heather, waxflower, and so on). Use just enough filler flower to accent the main flower, not overpower it. Position the filler blossoms above and behind the rose head or to the side, or lower in the front of the rose. Using floral tape, affix the stems of the filler flower onto the main flower stem.

Step 3 Add an ivy leaf (or other small broad leaf such as camellia, lemon leaf, or pittosporum) to the back or side of the design. Lacy foliages, such as leather leaf, ming fern, cedar, tree fern, or plumosa fern can be used as well. The foliage might have to be wired for additional support before taping it into the boutonniere (generally a stitch wire or wrap-around wire).

Step 4 Tape the boutonniere stem tightly and evenly using floral tape. A tight wrapping will secure the design and look nicer, without wires showing or poking through the tape.

Step 5 Curl the stem of the boutonniere forward around a pencil, if you wish, for an interesting, less blunt look.

Step 6 Attach a pin by inserting it through the thick portion of the wired stem. As shown in Figure 2-28, after making a carnation boutonniere, bend the carnation head forward. The flower will look nicer and fit more closely to the lapel.

2-28 *Pins should be inserted through the calyx before the boutonniere is worn. Remember to bend carnation flowers slightly forward to allow the boutonniere to lie flat on the lapel.*

2-29 *Two or three small flowers may be clustered together to form a boutonniere.*

Multiple-Flower Boutonniere

Once you have learned how to make a single-flower boutonniere, the design possibilities are endless. Two or three little flowers may be clustered together to form a unique boutonniere. It is important to remember to keep these designs fairly small. Avoid making boutonnieres that are so large and excessive that they look like corsages.

Step 1 Wire and tape two or three small flowers using floral tape.

Step 2 As shown in Figure 2-29, select the smallest of the flowers for the height of the boutonniere. Add the other flowers onto the stem of the first in a slight zigzag pattern. Angle the lower flower heads facing out.

Step 3 Add filler flowers for accent if desired. Add a few small broad leaves or tiny sprigs of foliage behind the main flowers.

Step 4 Use floral tape and complete the stems as desired. Add a pin.

Nestled Boutonniere

Smaller flowers (such as sweetheart roses, pixie buds, or a tiny cluster of statice, or other fillers) can be inserted into the center of a carnation. Knowledge of construction techniques for these unique, novelty boutonnieres is important, for they are often requested for proms and other special occasions (see Figure 2-30).

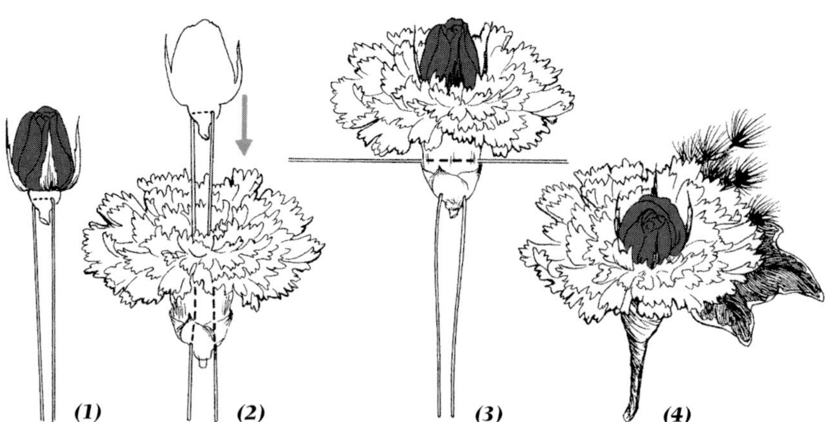

2-30 *For a nestled boutonniere: Step 1: Wire a rose and remove the ovary from the center of a carnation; Step 2: insert the ends of the wired rose through the top of the carnation down through the base of the calyx; Step 3: insert a wire into the carnation calyx and bend wires downward, keeping all four wires parallel (do not twist), and conceal wires with floral tape; Step 4: for the final step, add filler flowers and foliage.*

Step 1 Wire a small or sweetheart rose with a #22 wire. Do not tape.

Step 2 Select a carnation with a different color than the rose. Remove the pistil from the center of the carnation to allow you to nestle the rose further down into the carnation.

Step 3 Insert the ends of the wire through the top of the carnation head, down through the center, and out the base of the calyx. Position the rose down into the center of the carnation.

Step 4 Insert a wire (#24) through the carnation. Keep all wires parallel to each other. Tape the wires together using floral tape.

Step 5 Add filler flowers and foliage to the design if you desire. Add a pin.

Boutonniere and Corsage Stem Ends

The wired stems at the base of boutonnieres and corsages are generally visible, so it is important to cover all wires with floral tape. Often a distinctive corsage and especially a boutonniere may be set apart from all others by a simple twist or turn of the taped stems.

One of several simple techniques may be used for putting the final touch on the design piece itself. As shown in Figure 2-31, the simplest method is cutting the stem and leaving it straight. However, curled stems offer more unique options than does a straight stem. The stems may be easily curled by spiraling the stem around a pencil. The curl can be pulled to the side or stretched out in a number of ways.

Stems may also be given a more natural look by keeping individually wired and taped stems separated from one another, taping them together only at the top. For the design to appear as if the flowers have been plucked from the garden, vary the stem lengths.

2-31 *The simplest method to finish stems is to leave them straight; however, curled and clustered stems provide interesting options.*

Corsages

Corsages are worn by women on special occasions such as weddings, proms, and other formal events. Corsages are popular floral gifts during holidays such as Mother's Day, Easter, and Secretaries' Day, as well as many other times. Whether a corsage is worn by the mayor at a town fair, the woman giving the main address at a conference, the volunteer candy striper at the hospital, a woman on an anniversary dinner date, or a young girl graduating from elementary school, the corsage sets the wearer apart. Corsages and boutonnieres (see Figure 2-32) as floral gifts show appreciation and distinguish the wearer.

Corsages are commonly worn on the shoulder or the wrist. Smaller corsages may be worn in the hair, at the waist, or pinned to an evening purse. The style and fabric of the dress, current fashion trends, the occasion, and personal preference all dictate what type of corsage is preferred. And remember, the success of any design is dependent on proper and secure mechanics.

Single-Flower Corsages

Corsages with one main flower share similar steps of construction to a single-flower boutonniere. However, corsages differ from boutonnieres in the additions of bows or ribbon loops and accessories such as tulle and decorative novelties. The single flower may be a simple carnation or a more exotic gardenia, cymbidium orchid, or cattleya orchid (see Figure 2-33).

Step 1 Wire and tape an orchid (or other flower for a single-flower corsage).

Step 2 Prepare accessory items that are to be added into the corsage, such as two or three net (tulle) fans, one to three ribbon loops, or a tiny, unobtrusive bow. Lightly spray paint net fans, if desired, to provide color and accent.

Step 3 Prepare several leaves or sprigs of foliage and a few tiny sprigs of filler flowers.

Step 4 Place filler flowers behind and around the main flower as desired, and tape in position.

Step 5 Position foliage at the top and sides of the flower, and tape into place.

Step 6 Add net fans as background material around the perimeters of the corsage, if desired. Add ribbon loops for color and accent.

Step 7 Trim excess wire from the corsage as you add additional materials. Secure using floral tape and curl up the stem end. Add one or two corsage pins to the back of the corsage.

2-32 *Corsages and boutonnieres may be created in various flower combinations and styles for a wide variety of occasions.*

2-33 *Single-flower corsages are made more elaborate with the addition of filler flowers and foliage as well as a variety of accessory items such as ribbon loops, bows, and netting.*

Multiple-Flower Corsages

Corsages often have several small flowers grouped together, accented with foliage, ribbons, net tufts, and filler flowers. Constructing this type of corsage is similar to making several boutonnieres or single-flower corsages and then putting them all together into a secure, lightweight, visually lovely design. A variety of flowers, such as roses, feathered carnations, miniature carnations, chrysanthemums, stephanotis, and alstroemeria may be used. It is important to establish a theme and style for the corsage, so that harmony and unity are achieved. (Refer to Figure 2-34 to make a multiple-flower corsage.)

Step 1 Wire and tape three to seven small flowers. Choose a variety of flower types and sizes. Select some leaves or sprigs of foliage and filler flowers. Wire and tape for use in the corsage.

Step 2 Prepare accessory materials, such as net tufts, ribbon loops, and a bow.

Step 3 Begin the corsage with a small flower bud. Position a ribbon loop behind the flower and one in front. Tape these together. This flower will serve as the backbone or spine of the corsage.

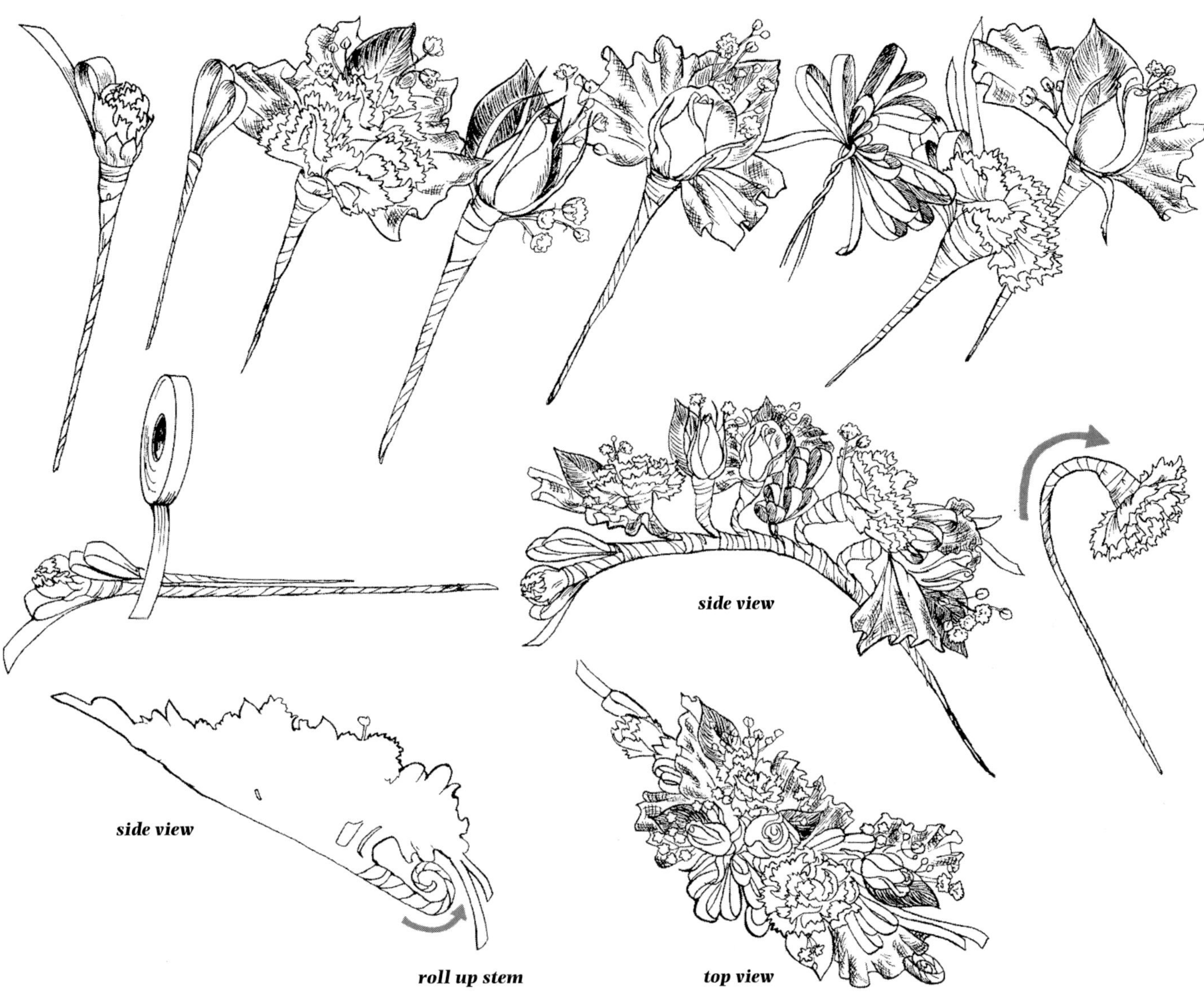

2-34 *Multiple-flower corsages are made by combining flower groupings and accessories into a harmonious and unified design.*

Step 4 Tape filler flowers, leaves, and accessory materials around and behind each of the flowers. Position individual flower sections (similar in appearance to boutonnieres and single-flower corsages) onto the main spine of the corsage in a zigzag pattern. Flower heads will begin to face out.

Step 5 If you wish to, add a bow in and among the flower sections. Make the bow integral to the design of the corsage so that it repeats the color and texture of the ribbon loops already taped into the design.

(*Continued*)

Step 6 To complete the visual flow of the design, position remaining flower sections below the bow so the flower heads are facing out and downward.

Step 7 Trim excess wire from the corsage as you tape in materials. Tape securely using floral tape and roll the stem end forward or to the side. The spine of the corsage should remain flat in order for the design to lie closely against the shoulder or the wrist.

Step 8 The corsage may be maneuvered into a variety of styles or shapes, such as crescent, round, vertical, or triangle, simply by bending and curving the wired stems into desired positions. Add two pins to the spine of the corsage.

Over-the-Shoulder Corsage

Over-the-shoulder corsages, sometimes called *epaulet designs,* are made to be worn on top of the shoulder and cascade down, both in front and back. Although these elegant, graceful designs are constructed similarly to other multiple-flower corsages, the smallest flowers on both ends must be wired with a fine gauge wire (#28 or #30 or finer), allowing them to cascade freely. These small flowers at the ends should be more widely spaced than the rest of the corsage, giving the corsage a more natural, graceful appearance.

Glamellia Corsage

A *glamellia corsage* is made from various sizes of gladiolus florets arranged in a way to resemble a camellia flower (hence the name for this corsage). The florets are removed from the stem, and the petals of the flowers are cut and wrapped around a gladiolus bud. These novelty corsages, though time consuming, are distinctively glamorous for special occasions (see Figure 2-35).

Step 1 Choose a slightly opened gladiolus bud. Remove the green calyx portions. Wire and wrap with floral tape.

Step 2 Cut the base of a floret, as shown Figure 2-35. The center portions of the flower will fall out, but the petals should remain intact. Position a section of petals against and around the center bud. Wrap a fine gauge wire around several times to secure the petals and then use floral tape. (Floral adhesive may also be used to attach the petals to the center bud).

Step 3 Add another section of petals circling around the other side of the bud. The glamellia will begin to take on a rounded shape.

Step 4 Continue adding sections of petals around the perimeters of the glamellia until the desired size is achieved. Wire and tape the base of the glamellia.

Step 5 Add several broad leaves underneath the glamellia to help support the newly formed flower and give it a more natural appearance. The glamellia flower may then be used in a corsage design.

2-35 A glamellia corsage made from gladiolus petals offers a distinctive and glamorus design option for a special occasion.

Wrist Corsage

Corsages may also be designed to be worn on the wrist rather than the shoulder. However, wrist corsages must remain lightweight, fairly small, and unobtrusive. Although made in the same way as shoulder corsages, these corsages must have a *wristlet* of some kind securely attached.

Several commercial wristlets or wristbands are available. Some have a band that is elasticized while others are plastic latch-type or velcro bands. The wristlet is attached to the back of the corsage with metal clamps (see Figure 2-36). For security, after pressing the clamps in place around the spine of the corsage, add floral tape around the clamps to keep them tightly in place.

Football Mum Corsage

Football mum corsages are traditional for high school and college homecoming football games, dances, parades, and other festivities. These corsages are regional; that is, in some areas they are considered the ultimate status symbol and in other areas they are virtually unknown.

Corsages are designed with one, two, or three standard incurve chrysanthemums (also called football mums) and a variety of long, cascading ribbons in the school's colors. Accessories such as miniature footballs, cowbells, megaphones, and tiny plush animals are popular additions to these designs.

2-36 *To attach a wristlet to a corsage, clamp down the metal brackets onto the back of the design and secure them with floral tape.*

2-37 *Football mum corsages are traditionally worn for high school and home coming events in some regions of the country. They are made with standard incurve mums and highly decorated with accessories. Photo courtesy of Florists' Review Enterprises.*

Often the first initial of the school name is represented on top of one of the mums with the name of the individual wearing the corsage spelled out with glitter or stick-on letters on one of the flowing ribbons (see Figure 2-37).

Because of the size and weight of these corsages, mechanics are all important and everything must be wired and glued securely. To keep petals from *shattering*, mums must be sprayed on the backside with a special clear glue. The following steps of construction may be altered for making single and simple, or large and lavish designs.

Step 1 Choose a standard incurve chrysanthemum (football mum) free of blemishes. Remove all but one inch of stem below the flower head. Hook-wire the flower with a gauge #22 wire. (Alternate wiring techniques such as pierce or insertion may be used.) Tape the wired stem using floral tape. If desired, wire and tape a second or a third football mum, depending on the style and size of design.

Step 2 Spray the flower head with a spray sealer, which will prevent petals from shattering.

Step 3 Stitch-wire camellia or salal leaves with gauge #24 wire. These wired leaves will help to support the otherwise fragile mum. Tape the stems of the wired leaves using floral tape. Place these leaves around the mum to add background while giving support to the flower. Leave all wires parallel to one another. Tape the wires tightly together.

Step 4 Long ribbon streamers and a large bow may be added at the base of the design. Streamers may be braided, knotted, or twisted.

Step 5 Accessories such as dangling miniature footballs or cowbells may be added to the design. Letters and numbers may be made out of chenille stems; these may then be glued onto the top of the flower head or attached with a wire.

Step 6 Securely tape the stem. Make sure all accessories are secure. The entire design may be sprayed with a light coat of glitter or sealer spray.

Other Floral Designs to Wear

Floral pieces are often worn in the hair, attached to a hat, pinned to a purse, or worn as a lei. As in other designs, the occasion, clothing, individual preference, and fashion trends all help determine the desired design. As with wrist corsages, keep designs lightweight, with an appropriate shape and size for the setting.

Flowers for the Hair

Small floral pieces are often designed for individuals to wear in their hair for special occasions such as proms and weddings. Tiny flowers or filler flower clusters may be secured in the hair with hairpins. Small floral designs (similar to boutonnieres) may be attached to a barrette, comb, or hair clip with wire or glue (see Figure 2-38).

Chaplet

A chaplet is a floral wreath or garland for the head. These designs are popular for special occasions such as proms and weddings (Figure 2-39a). Since floral wreaths are often worn by young flower girls at weddings, it is important that head measurements be taken before making these designs to ensure a proper fit.

2-38 *When attaching a floral piece to a bar-rette, secure it with wire or glue.*

2-39a *Floral head wreaths, often called chaplets, are popular for weddings, dances, and other special occasions. Photo courtesy of The John Henry Company.*

Step 1 Measure the head. Cut a #22-gauge wire to the length needed, adding on an extra inch to allow for hooking the wire ends together to form a wreath (using floral tape, two straight wires may be taped end to end to form a longer piece, or spool wire may be used).

Step 2 Tape the wire, as shown in Figure 2-39b, using floral tape.

Step 3 Wrap strands of ivy around the wire. Secure the ivy periodically with floral tape. Or, another common procedure for adding foliage is to bind tiny clusters of foliage, such as ivy or leather leaf, with fine wire (#28); each additional cluster of foliage should cover the stems of the previous cluster. Floral tape will secure foliage clusters in place.

Step 4 Add tiny flowers and fillers in a random or orderly pattern with wire and floral tape.

Step 5 Form the wire into a ring and insert the straight wire end into the opposite hooked end, as shown.

Step 6 If you wish, add delicate ribbon streamers and a tiny bow in the back where the garland comes together, concealing the final mechanics of the chaplet.

2-39b *Although time consuming, floral chaplets are easily made with wire, tape, and sprigs of flowers and foliage. Because they are often worn by young girls, it is important that head measurements be taken before making, for proper fit.*

Lei

The *lei* originates in Hawaii and is a garland or wreath of flowers and leaves, generally worn around the shoulders about the neck. Leis vary greatly according to the flowers, foliage, and the manner of assembly. For example, a simple lei may be made with carnations (see Figure 2-40).

Step 1 Gather forty-five to fifty-five carnations and remove their stems.

Step 2 Using a special medium-weight lei needle (or long darning needle), thread dental floss, fishing line, or lei string (about 7 feet long, doubled to $3\frac{1}{2}$ feet long or 42 inches) into the needle.

Step 3 To make each carnation fuller, gently brush across the petal tips of each flower, as shown.

(*Continued*)

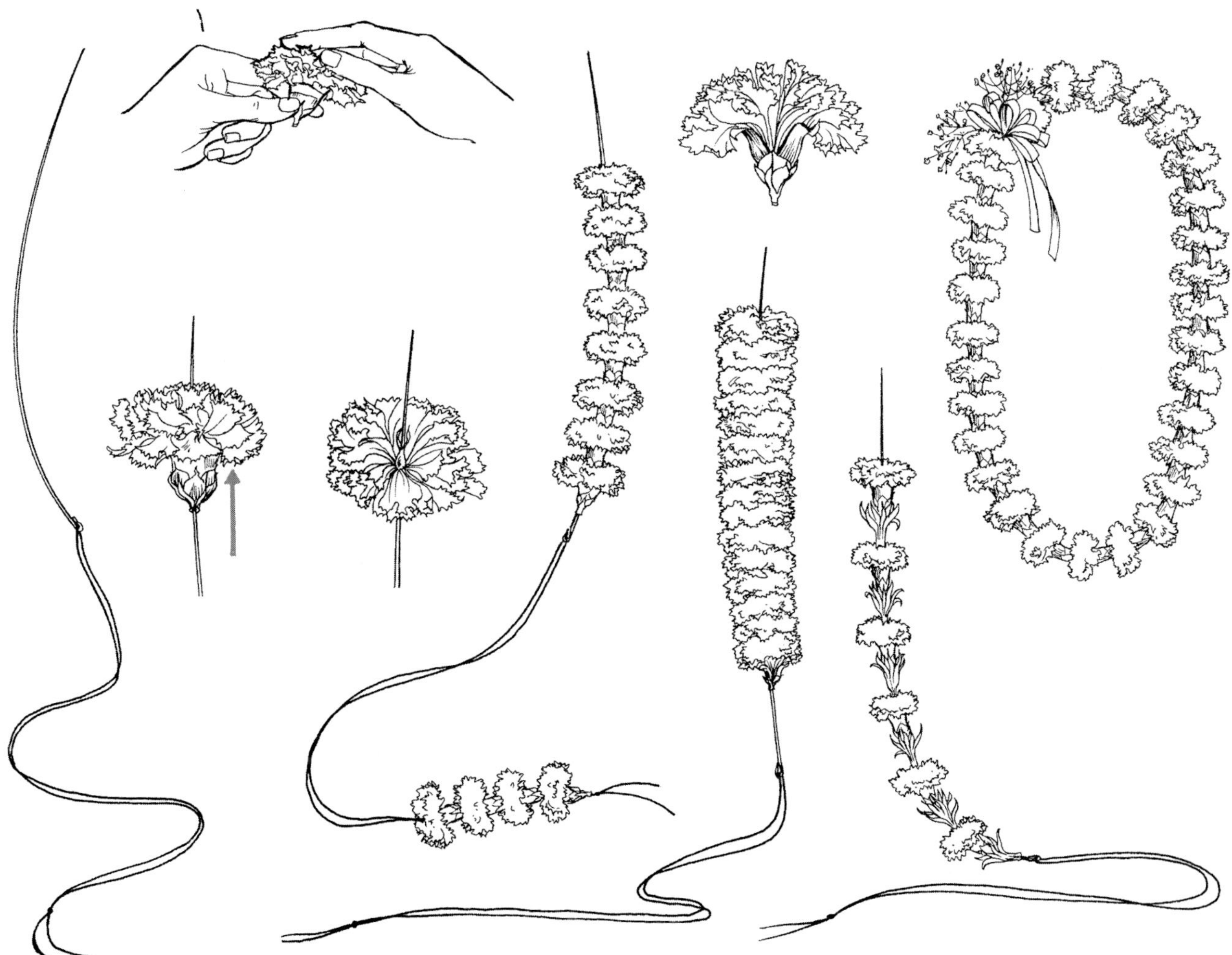

2-40 *A lei, or wreath of flowers, is worn around the shoulders about the neck. Lei styles vary greatly according to flowers, colors, and alternating patterns (shown here are a "single" carnation lei, a "double" carnation lei, and a lei designed with carnations and tuberose blossoms).*

Step 4 Begin threading a carnation onto the string by first inserting the needle into the calyx. Continue pushing the needle through the calyx until it goes through the seed pod and comes out through the center of the flower. Thread another carnation onto the string in the same fashion. The carnation flowers should fit snugly next to one another, and the green calyx should be visible between each carnation head.

Step 5 Thread enough carnations to make the lei of the desired length. Remove the lei needle and tie the string ends together. The connecting point may be concealed with a bow.

A lei with a "double" carnation pattern is made in much the same way as a "single" carnation lei.

Step 1 Gather twenty-five to thirty-five carnations and remove their stems.

Step 2 Using a special medium-weight lei needle (or long darning needle), thread dental floss, fishing line, or lei string (about 7 feet long, doubled to about $3\frac{1}{2}$ feet long or 42 inches) into the needle.

Step 3 Split carnation sepals (calyx) as shown, keeping the flower intact, to allow the carnation flower to spread out flat, forming a wider, thicker lei.

Step 4 Begin threading a carnation onto the string by first inserting the needle into the calyx. Continue pushing the needle through the calyx until it goes through the seed pod and comes out through the center of the flower. Feed carnations onto lei string with the green calyx tucked into the center of the previous carnation to produce a thicker, fuller pattern. No calyx should be visible.

Step 5 Thread enough carnations to make the lei of the desired length. Remove the lei needle and tie the string ends together.

Flowers to Hold

Handheld flowers and bouquets are often chosen by young ladies to enjoy at special events. The custom of carrying flowers has been a romantic and sentimental tradition since the English-Georgian and Victorian times. Miniature nosegays are sometimes popular for proms and other school dances and events. It is important to make these designs lightweight and easy to carry. Flowers may be inserted into a wet foam holder or they may be tied together, as in a hand-tied bouquet.

Sealers

Aerosol and liquid *sealers*, often referred to as finishing sprays and dips, may be used to seal the porous surfaces of flowers and foliage. Sealers inhibit water loss, helping flowers and leaves in corsages, boutonnieres, and other floral pieces to remain firm for a long period of time. Sealers should only be applied to firm, healthy flowers and foliage; they will not help flowers and foliages that are already wilted. After spraying a floral piece, allow the sealer time to dry before packaging the design.

Packaging

Corsages, boutonnieres, and other floral pieces for special occasions should be packaged carefully to prevent moisture loss, protect the floral design, and provide an attractive presentation for the receiver. Several types and sizes of bags and boxes are available for packaging flowers that will be worn.

Step 1 Place the floral piece on a layer of shredded wax paper (often called orchid grass). The *orchid grass* will cushion and help protect the design while it is packaged.

Step 2 Gently place the floral design and orchid grass into an appropriately sized bag. Fold the bag and close with boutonniere or corsage pins (or if pins are already attached to the design, staple the bag shut). It is not always necessary to use a bag within clear plastic boxes or boxes with windows; simply place the orchid grass in the bottom of the see-through box and place the floral piece on top.

Step 3 Place the floral design (within the bag) into an appropriately sized box. A ribbon in a corresponding color may be tied in a pretty bow around the box.

Flowers to wear are special accessory items that may be designed in an infinite number of ways. Many special occasions and events are traditionally associated with the wearing of flowers such as weddings, proms, homecomings, Mother's Day, and many other formal or sentimental times. Floral pieces to wear must be designed in the appropriate shape, size, and style to fit the purpose and place for which they are intended to be worn. Proper wiring and construction techniques are essential to keep these fresh designs secure and long lasting. Knowing and practicing basic techniques of construction for boutonnieres, corsages, and other floral pieces to wear will give you confidence in your own ability to create truly distinctive designs.

Contemporary Design Styles and Techniques

Often misinterpreted as weird, bizarre, and undesirable, *contemporary designs* are those that are currently in fashion, popular, and representative of leading trends in creativity (see Figure 3-1). Contemporary design encompasses *classic design* or *traditional naturalistic design, linear design, modernistic design,* and *experimental design.* Often, certain styles of arrangement, such as experimental designs, are considered contemporary and advanced because they are new, nontraditional, and unfamiliar. This section focuses on contemporary designs that are complex or advanced in nature, requiring specialized skills, mechanics, and techniques. Experienced, creative floral designers are interested in advanced and contemporary arrangement styles because these designs offer distinctive artistic alternatives. *Advanced design* provides a creative challenge and requires floral designers to keep current in their knowledge of design styles and techniques.

Classic Design Styles

Classic floral arrangements are generally mass bouquet designs that continue in popularity and are in fashion due to their simplicity and elegance. Often called *traditional,* classic floral designs are versatile and may be displayed in a variety of ways. Traditional mass floral arrangements include circular, oval, triangular, and fan-shaped bouquets. Advanced classic styles include *mille fleurs design, Biedermeier design, phoenix design,* and *waterfall design.*

3-1 *This striking, contemporary design featuring miniature heliconia, gerbera, and banksia demonstrates dynamic relationships between textures, colors, and forms. A qualified contemporary designer has the confidence to break away from the expected. From Design with Flowers © 1991 Herbert E. Mitchell; photography: deGennaro Associates, Los Angeles, CA.*

Mille Fleurs Design

The arrangement style *mille fleurs*, sometimes referred to as *mille de fleurs*, means "a thousand flowers" and refers to having an all-over multicolored pattern of many flowers. Having emerged in the mid-19th century in Europe, this traditional style of arrangement incorporates many different flowers and colors. Generally fan-shaped or rounded, these designs express opulence and abundance with the many varieties of brightly colored flowers juxtaposed in an arrangement (see Figure 3-2).

Step 1 For these traditional designs, choose a container to support the physical as well as visual weight of the vast number of flowers that will be used.

Step 2 If floral foam is used to hold all of the flowers in place, secure the foam in the container. (Do not use floral foam if you select a clear vase.) Without the use of floral foam, a natural grid will soon develop as a tall vase is filled with flowers by *lacing* the stems. The multiple stems secure each other, and the placement and angling of the flowers becomes more fixed and distinct.

3-2 *A mille fleurs, "a thousand flowers," design contains many different flowers and colors in a multicolored pattern. This design is a contemporary and creative interpretation of the classic, tight, and structured* mille fleurs *arrangement. From* Design with Flowers *© 1991 Herbert E. Mitchell; photography: deGennaro Associates, Los Angeles, CA.*

Step 3 Choose a great variety of flower types and colors in order to achieve the "thousand flowers" appearance. Begin by adding taller, line flowers in a radiating pattern.

Step 4 Next, add rounded, mass flowers within the boundaries of the line flowers to achieve fullness and the desired colorful pattern.

Step 5 Add the final placements, whether they are line, mass, form, or filler flowers, to provide forms, colors, and textures in certain parts of the arrangement to give an all-over floral pattern. Remember, the flowers should be arranged in a random, loose, and airy fashion to provide a feeling of natural opulence.

Biedermeier Design

The *Biedermeier* style originated in Austria and Germany during the post-war years 1815 to 1848. It is associated with a heavy style of furniture, similar to French Empire and English regency designs.

A *Biedermeier* floral arrangement is generally compact, rounded, or slightly conical in shape and consists of concentric rings of flowers as shown in Figure 3-3. Each circular row is composed of the same flower. Each floral ring contrasts to adjacent rows of materials. The contrast of colors, forms, and textures with each row promotes visual interest.

3-3 *The* Biedermeier *floral arrangement is compact, rounded, or slightly conical in shape and consists of concentric rings. Carnations, sweet William, and roses exemplify this classic style of design. From* Design with Flowers *© 1991 Herbert E. Mitchell; photography: deGennaro Associates, Los Angeles, CA.*

3-4 Generally filling a compote container, the Biedermeier floral design consists of concentric circles of foliage and flowers. Starting at the rim of the container, insert the same type of leaves or flowers in tight rows. A rose or other fragrant flower is usually placed at the top, completing the design.

Many adaptations of the *Biedermeier* style are possible with spiral patterns or looser, mixed flower placements, still in planned alternating rows. Often berries, leaves, nuts, small vegetables, and other materials are placed in concentric rings alternating with rows of flowers, creating further contrast and interest.

Step 1 Select a compote or other container. Fill with floral foam. Next, with a knife, contour the foam's edges into a rounded, pyramidal, or slightly conical shape. Secure the foam in the container with waterproof tape.

Step 2 Begin by placing a ring of leaves (such as salal, galax, or small pieces of leatherleaf) into the floral foam outward, over the rim of the container (see Figure 3-4).

Step 3 Next, add a ring of flowers above the row of leaves. Cut flower stems to about two inches and insert the stems into the foam until the flower heads gently rest against the surface of the foam.

Step 4 Continue adding concentric rings of contrasting flower types, colors, and textures. The rows of flowers must be close together to hide the floral foam and tape. Filler flowers, such as tiny sprigs of babies breath or statice, may be added in between concentric rings already in place to help conceal any mechanics and to create additional contrasts between rows.

Step 5 Complete the design by placing a single flower, commonly a fragrant rose, at the top of the design.

Phoenix Design

The inspiration and name for this style of design comes from the ancient Egyptian mythological bird, the phoenix (see Figure 3-5). Legend tells of the lone phoenix that lived in the Arabian Desert for 500 or 600 years and then set itself on fire. It then rose renewed from its ashes to begin another long life. The Egyptian bird is a symbol of renewal and immortality.

A phoenix-style floral design has a base of flowers in a traditional, rounded, compact shape. Bursting from its center rise tall, flowering branches or line flowers, which represent renewal and strength. This distinctive design is often used as a party centerpiece or home decoration.

3-5 Legend tells of the ancient Egyptian, mythological phoenix bird that rises from fire and ashes to begin a new life.

Step 1 Select a compote or other container and fill with floral foam that extends slightly above the container rim. Contour the foam's edges with a knife. Secure the foam in place with waterproof tape.

3-6 After securing foam into the container, first insert tall stems deep into the center of the foam. Position the tall stems so they radiate outward slightly at the top. Complete the design by inserting flowers and foliage at the base to form a circular arrangement.

Step 2 As shown in Figure 3-6, insert tall, flowering branches, line flowers, or other linear material into the center of the foam. These stems should radiate out at the top.

Step 3 Next, insert flowers and foliage at the base of the design to form a traditional round, compact arrangement. Place the flowers at the top of this round design close to the base of the flowering branches in order to hide floral foam and blend with the rising branches.

Waterfall Design

The waterfall design, representative of a waterfall (see Figure 3-7), can be traced back to the early 1900s. Initially created for bridal bouquets in Europe, this pendulous style of floral arrangement, often depicted in paintings of the *Art Nouveau* period, has been rediscovered and become a popular container design. Romantic and naturalistic, these flowing designs are a contemporary version of the traditional floral cascade.

This style is characterized by a downward flow of materials, often heavy with foliage. Nonbotanical elements, such as feathers and yarn, are frequently added to give an untidy, undisciplined appearance. Representative of splashing, glistening water, reflective materials such as small fragments of mirror and metallic thread are often incorporated into the bouquet to give the appearance of splashing sunlight. Flowers and foliage that are long, trailing, and pliable are essential to form the downward cascade. Bear grass, sprengeri fern, plumosa fern, conifers, vines, ivies, twigs, and string smilax are examples of useful materials for creating the long, curving form of the waterfall design. The materials flow from the center of the design out and over the container edges.

3-7 A cascading waterfall gives inspiration for the waterfall floral design.

Depth is created through the layering of materials. Layers of foliage alternate with layers of flowers. Colors and textures also alternate within the design, displaying diversity in materials. The waterfall must have adequate room to cascade downward so a tall container is generally needed. However, this design style is also effective flowing over the edge of a pedestal, table, mantle, or shelf. Designs that are elevated provide dramatic results.

Step 1 Select a pedestal vase, urn, or other tall container. Secure floral foam in the container. Foam must extend several inches above the container rim to allow horizontal and downward stem positioning.

Step 2 Select flowers and foliage that have long and flowing curves. Begin the design by placing long, curving foliage to form a cascade on one side of the design. (Waterfall designs may be constructed in different ways. In asymmetrical designs, the materials cascade predominantly on the side of the container whereas in symmetrical designs, the materials cascade all the way around the container, displaying equal balance.)

Step 3 Place additional foliage around the rim of the container, extending it downward (see Figure 3-8). The side opposite the long cascade may have a much shorter cascade to help balance the design visually.

Step 4 Add layers of flowers; some should face outward and others should face downward. Cover the

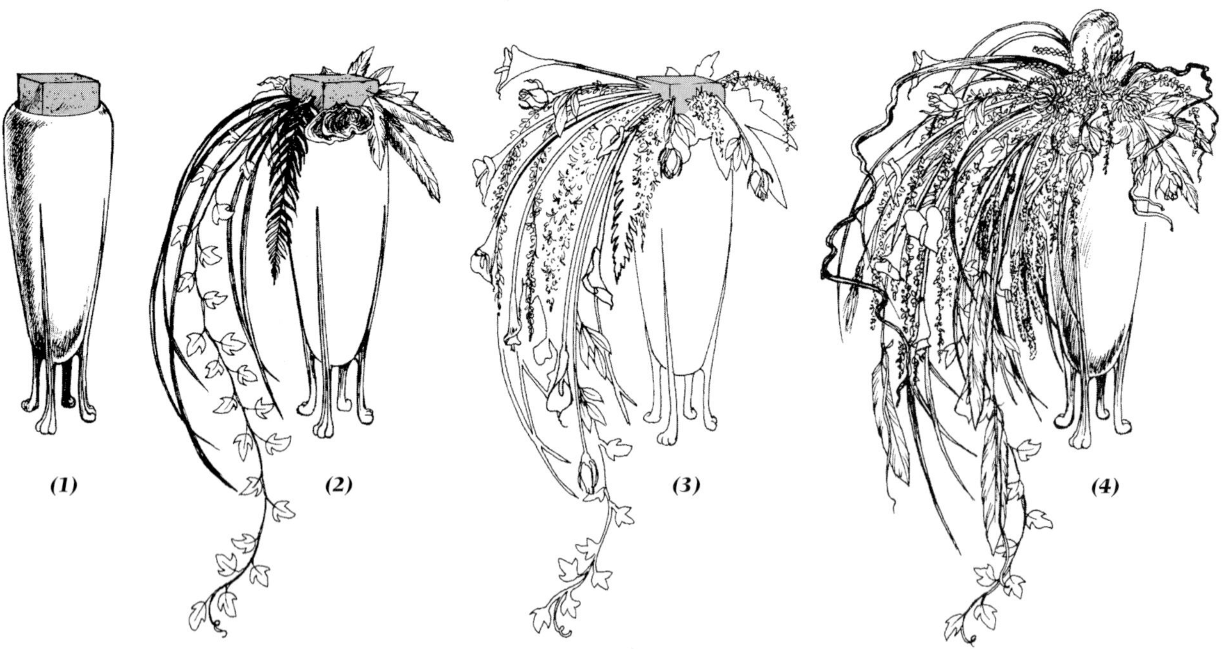

3-8 *Waterfall design—steps of construction: Step 1. Select a compote or other tall container. Secure floral foam above the container rim to allow for downward positioning of stems. Step 2. Place long, curving foliage deep into the foam to form a flowing cascade down one side of the container. Next, place shorter foliage around the rim of the container. Step 3. Add layers of flowers flowing down the foliage cascade. Next, add other, shorter flowers around the container rim. Flowers should face outward in all directions, with a gradual transition downward in the cascaded area. Step 4. Cover floral foam with more layers of foliage and flowers. Nonfloral reflective materials may be added to the cascade to give a shimmering appearance.*

central base of the design with short-stemmed flowers and foliage. Continue adding layers of flowers and foliage that allow the central portion of the design to flow into the cascading areas. Nonfloral, reflective materials may be added to give the design the image of shimmering light. The waterfall should seem to flow from the center of the container; however, some lines may cross to add interest. Arrange materials so that the design will be pleasing from all sides and rich in depth and texture.

Naturalistic Design Styles

Often appearing wild and uncultivated, floral designs termed *naturalistic* or *natural* are based on nature. Natural designs do not appear contrived or artificial, but represent a slice of the outdoors. These designs emphasize the beauty of flowers without manipulation. Containers must harmonize with the flowers and other materials in the design. *Botanical, vegetative,* and *landscape designs* all reflect some aspect of nature.

Botanical Design

Botanical design is considered a new and contemporary American floral arrangement style. This design represents nature in the study of the life of a plant through the close-up look of a bulb flower. Generally focusing on one kind of bulb flower, the parts of the plant, including the buds, blossoms, foliage, stems, bulb, and roots are visible. The plant's natural environment is often depicted in the design's base with stones, mosses, and other bulbed flowers.

Step 1 Select a low container or basket. It should be wide and deep enough to allow for insertion of small live plants (in their soil and containers) if desired. Cut a block of floral foam, contour the edges, and place it into the container.

Step 2 To show the bulb and roots of the plant, it is necessary to secure the bulbed flower into the foam with a wooden pick, as shown in Figure 3-9. Insert the sharp end of the pick into the bulb, and insert the opposite end of the pick securely into the floral foam. Add more of the same type of flower into the foam adjacent to the bulb and roots to make a natural clustering of flowers.

Step 3 Next, add other bulbed flowers and potted plants to the design to form a natural setting. These flowers and plants should be subordinate to the main bulbed flowers that show the bulbs and roots.

Step 4 Conceal the mechanics (such as flower pots and floral foam) with stones, mosses, and twigs. The base should form a natural-looking environment.

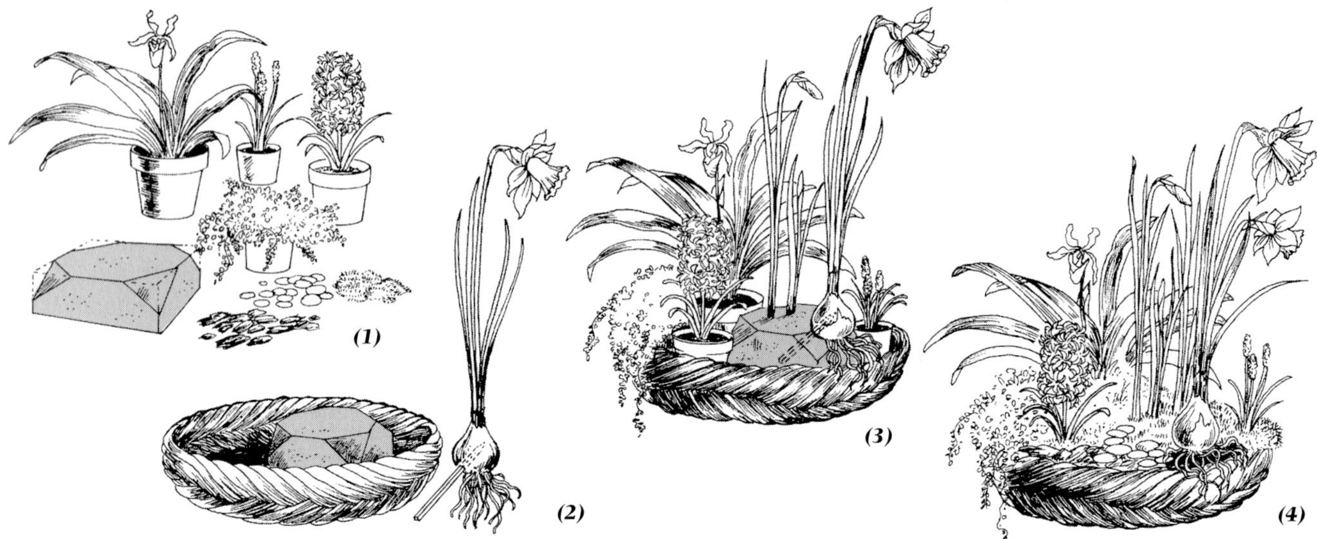

3-9 Botanical design—steps of construction: Step 1. Select a low container or basket. Select plants and other materials to form a natural-looking environment. Step 2. Contour edges of the floral foam. Insert a sharp wooden pick or other mechanical aid diagonally into the bulb to allow the essential bulb and roots to be visible in the design. Step 3. Place materials into the container in a natural, clustered pattern. Step 4. Complete the design by covering the foundation area with mosses, stones, and other materials.

Vegetative Design

Vegetative design presents plants as they grow in nature (see Figure 3-10). This natural design simulates a small slice of nature. Flowers and other materials are arranged in a container as they might be found in a natural setting. (Taller growing flowers are placed high in the design, and shorter growing flowers are arranged low in the design.) Flowers and foliage are selected according to seasonal compatibility. Plants that grow together in nature are juxtaposed with one another in a parallel or radial style. These designs should have visual interest on all sides and may easily be used as a centerpiece.

3-10 Vegetative designs present plants juxtaposed as they grow in nature.

Step 1 Select a low container. Secure floral foam in the container. Contour the foam edges to provide a less rigid block shape.

Step 2 Work from the top of the design downward. Do not place the tallest flowers in the center of the design. Arrange them off center for a more natural appearance. Arrange materials on all sides.

Step 3 Because vegetative designs are a glimpse of the outdoors, do not alter flowers, buds, leaves, or stems. Leave flowers as they would be found outside. Blemishes, mature blossoms, weeds, and thorns remain in these designs.

Step 4 Layer the heights of blossoms and alternate textures and colors. Bunch similar materials together.

Step 5 Complete the design by adding mosses, rocks, twigs, clumps of grass, and other materials compatible with the flowers and season.

3-11 *Landscape designs depict a larger panoramic view of a natural setting or landscape than do vegetative designs.*

Landscape Design

Although appearing similar to the vegetative style, a landscape design arrangement depicts a larger area of nature (see Figure 3-11). Flowers, branches, and foliage may represent parts of a natural landscape or a groomed garden. Trees, bushes, flowers, and the ground are all represented and organized in color groupings. Patches of foliage, rocks, bark, sand, and moss are placed as they would be found in nature at the design's base to give visual relief from the groups of color. Like the vegetative design, all materials selected must grow in the same environment and during the same season. In contrast to all-sided vegetative designs, landscape arrangements are generally one-sided.

Step 1 Because landscape designs represent a larger, panoramic view of nature, often it is necessary to select a large and low rectangular, oval, or rounded container. Secure floral foam in the container.

Step 2 Place taller materials in the back of the design. These branches represent tall, far-off trees. Materials are generally grouped with contrasts in colors and textures, the same way plants are often found in nature. Do not arrange materials symmetrically; asymmetrical positioning is more natural.

Step 3 Complete the design by placing moss, rocks, twigs, and other materials at the base.

Linear Design Styles

Contemporary designs that are termed "linear" emphasize line and visual movement. Clean, taut lines combined with essential negative space make these designs distinctive and impressive. Form, proportion, and rhythm are necessary principles of design in linear arrangements. Distinctive, advanced linear styles include *western line design, parallel systems, new convention,* and *formal linear.*

Western Line Design

"Western line" is a general term for both symmetrical and asymmetrical triangular arrangements. Typically, the contemporary application of a *western* line design is an open triangle with a focal point near the base from which all stems radiate. The framework, height, and width are all patterned similarly to a scalene triangle. As shown in Figure 3-12, a Western line design is an open and striking arrangement. The height of the design is generally at least one and a half to two times the width of the arrangement. These arrangements have a prominent vertical line with an opposite downward sweeping line. It is important not to fill in the body of the arrangement because the open, negative space is essential to give the design visual distinction. The height of the design is often exaggerated while the stems creating the width plummet downward.

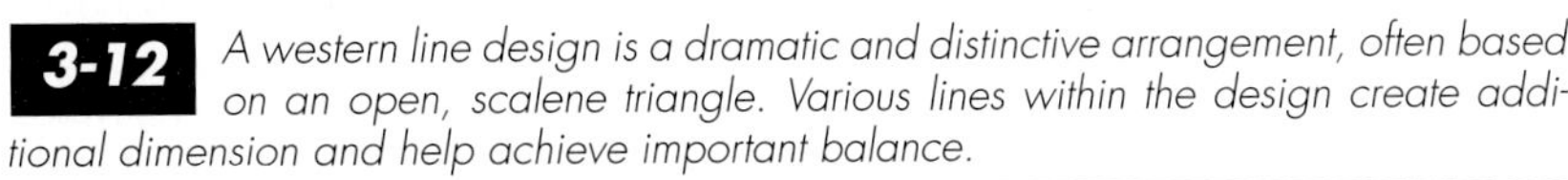

3-12 *A western line design is a dramatic and distinctive arrangement, often based on an open, scalene triangle. Various lines within the design create additional dimension and help achieve important balance.*

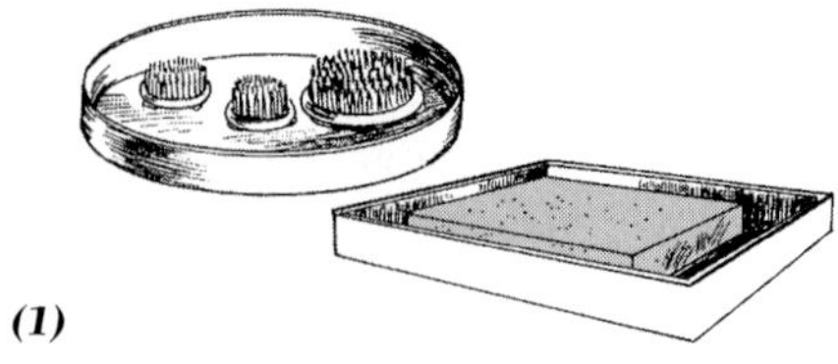

(1)

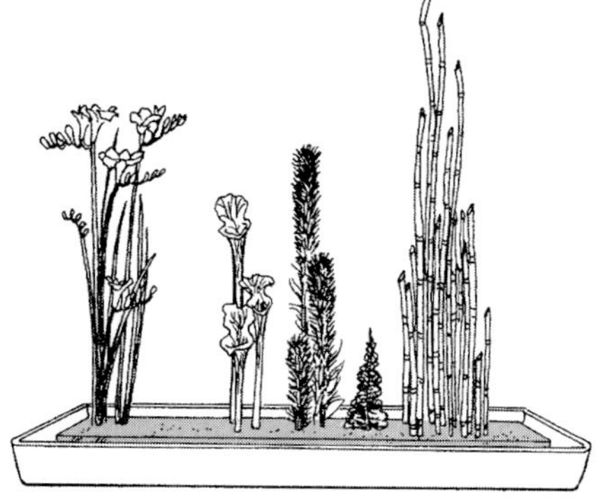

(2)

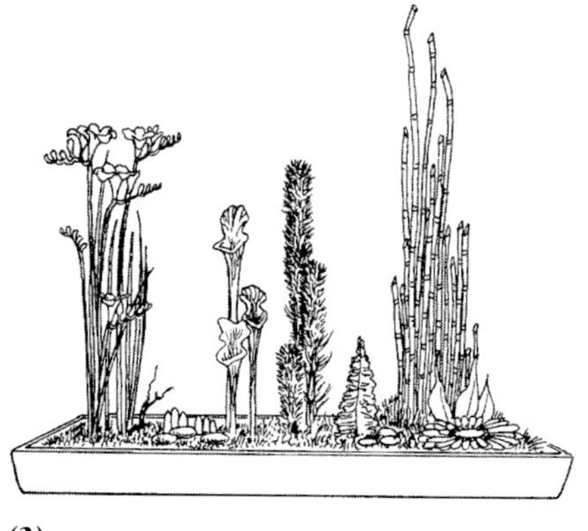

(3)

3-13 *Parallel systems design—steps of construction: Step 1. Select a simple, low container. Secure a foundation area of needlepoint holders or foam. Step 2. Choose flowers and foliage with clean, linear stems. Group similar flowers and foliage and allow plenty of negative space between differing materials. Step 3. Complete the design using a variety of basing techniques.*

Parallel Systems Design

A *parallel systems* arrangement consists of clusters or groups of flowers and foliage (see Figure 3-13). Developed by European floral designers in the late 1980s, this linear style has become a popular alternative to mass style designs. Each group in this design style consists of one type of flower or greenery. Generally, the plant material is set in a vertical pattern with negative space between each section. The negative or empty spaces allow the eye to travel through the arrangement.

This style of design uses the technique of *parallelism* in which all stem placements in each group are parallel to each other. The materials within each group may be repeated several times to increase the height and width of each cluster. Positioning flowers in a parallel manner, or parallelism, is a technique that may be used in a number of different styles such as vegetative, landscape, new convention, and abstract design.

The composition's base is generally covered with groups of leaves, mosses, stones, and other materials to form a decorative or vegetative look. Or the base may be left clean, with perhaps just a pool of water or a bed of stones, appearing formal and "architectural." All materials in the design should stay within the container edges. The container should be simple in design. Most often the container is low and rectangular or oval, but other shapes may be used. A parallel systems design displays open balance and is not prominently asymmetrical or symmetrical, but instead relies on a perceived balance.

Although unique and distinctive, parallel systems arrangements are versatile in their use. Because of the open areas within these designs, they work well as centerpieces. Also, they may be made on a grander scale for use in large areas.

Step 1 Select a simple, low container. Secure a foundation area in the container with needlepoint holders or floral foam.

Step 2 Choose flowers and foliage that have a clean, linear appearance to form the various groups. Visually divide the design in sections and establish groups within the composition. Each grouping should be made from one type of flower or foliage. Remember to allow adequate negative space between each cluster. It is best to stagger the heights and spacing of the various groups.

Step 3 Complete the design by concealing the mechanics. This step may be accomplished by using a variety of basing techniques, such as *clustering, pillowing, pavéing, layering,* and *terracing.* The design should appear neat and organized.

New Convention

Although similar to the parallel systems arrangement, *new convention* designs not only incorporate vertical groupings of plant material, but horizontal groupings as well. The vertical flower and foliage clusters are repeated low in the design. These horizontal lines are placed at sharp right angles, at the base of the vertical group they reflect. The horizontal groups form linear extensions to the front, back, and sides and are often made in the same materials or similar colors to the adjacent vertical clusters. The

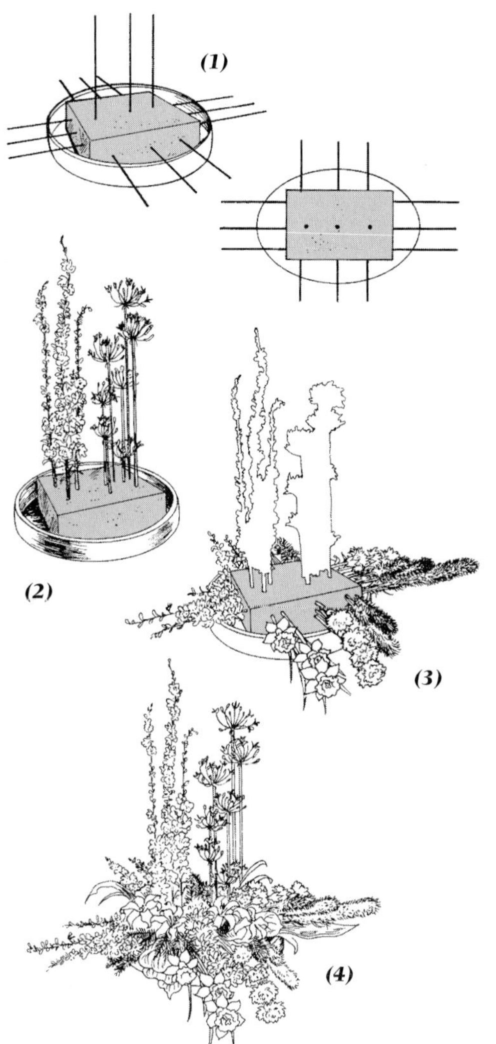

3-14 *New convention design—steps of construction: Step 1. Select a low container. Plan the vertical lines and horizontal lines before construction. When viewed from above, the horizontal lines extend outward at 90-degree angles to the sides, front, and back. Step 2. Begin the design, similar to a parallel systems arrangement, by placing groups of material in a vertical, parallel pattern. Step 3. Next, insert stems into the sides of the floral foam. These horizontal groups should be placed at right angles to the vertical groupings. Horizontal extensions are generally reflections of the vertical flowers and colors. Step 4. Complete the design by using a variety of basing techniques.*

horizontal lines are shorter than the vertical lines, using less material. Not all the vertical groups need to be reflected in the horizontal positioning at the base. Negative space must exist between the vertical sections and the horizontal extensions. The combination of vertical groups juxtaposed with horizontal groups results in distinct visual strength and drama.

Containers for these unique designs are generally low and rectangular, but other shapes may be effective. Several basing techniques such as layering, terracing, and pavéing with mosses, leaves, and flower heads are often patterned in rows, parallel to the horizontal stems.

Step 1 Select a low, rectangular or rounded container. Secure floral foam in the container. The foam must extend above the container rim to allow for horizontal stems. Begin by visually sectioning and planning the design (see Figure 3-14).

Step 2 Place linear materials to form the vertical groupings. Stagger stem heights within each group. The various clusters should also be at different heights.

Step 3 Next, insert stems into the sides of the floral foam, extending at right angles to the vertical groups. The horizontal lines should be shorter than the vertical groups and extend out in the front, back, and sides over the container rim.

Step 4 Complete the base of the design with short flower heads, mosses, and leaves. Use a variety of basing techniques such as layering, terracing, and clustering to provide textural variety and visual interest.

Formal Linear

As the name suggests, forms and lines are dominant in the *formal linear* style. Often referred to as *high style* design or *Art Deco*, these asymmetrical arrangements emphasize shapes, angles, and clean lines. Through the use of minimum shapes and quantities, and the use of negative space, the beauty of the flowers, foliage, and stems are accentuated. Generally, similar materials are grouped to emphasize shape, line, color, and texture. Lines may be vertical, horizontal, diagonal, or curved. Adjacent textures are contrasted, heightening their visual and tactile differences. Often created with exotic tropical flowers and foliage, this style of design highlights their unique shapes, colors, patterns, and textures. The entire design must be kept neat and organized with clean lines. The concept of "less is more" is essential to this distinctive design (see Figure 3-15).

Step 1 Select a container that will appropriately present the materials and style of design. Secure floral foam in the container. Floral foam should extend above the container rim to allow for horizontal and downward positioning of stems.

(Continued)

3-15 *Formal linear design. Left: Select a tall container. Extend floral foam above container rim. Select linear materials to establish the height of the design. Place materials in groups. Right: Complete the design by filling the central area and concealing floral foam. Use a variety of basing techniques.*

Step 2 Select linear materials to establish the height of the design. Group similar flower types.

Step 3 Next, add other groups of flowers and foliage. Allow plenty of space between the various groups in the design.

Step 4 Complete the design by filling in the central area of the design and concealing any floral foam. A variety of basing techniques may be used, such as clustering, layering, and terracing.

Containers are often tall but may be low. A variety of vase shapes are effective for the formal linear style. The container should reflect the high style feeling, allowing it to become an integral part of the entire design.

Modernistic Design Styles

Modern contemporary trends in floral arrangement are often termed modernistic designs. These designs may reflect contemporary fashion, colors, and even attitudes. Often termed experimental, modernistic designs are usually trendy and faddish; yet, for a time they may fill a certain need. Unique modernistic floral styles include *sheltered design, pavé, new wave,* and abstract design.

Sheltered Design

A *sheltered design* is "protected" within the container. Often, all materials are arranged below the container rim and may only be viewed by peering down inside. These designs, although seeming less dramatic and showy than most styles, require a closer, sometimes longer look. Slower-paced, sheltered designs offer an artistic alternative of privacy and protection from the outside world.

Step 1 Select a container that will allow you to arrange materials within its edges. Secure floral foam with an anchor pin or use a needlepoint holder (see Figure 3-16).

Step 2 Group similar flowers and leaves. Cut stems short, allowing the flower heads to rest on top of the foam. Use a variety of basing techniques such as clustering, grouping, and pavéing.

Step 3 Complete the design by concealing any floral foam. Stones, mosses, sand, twigs, and other materials may be added to the base of the container around the flowers and buds. Often, a shallow pool of water with a few stones will give the entire arrangement a restful appearance.

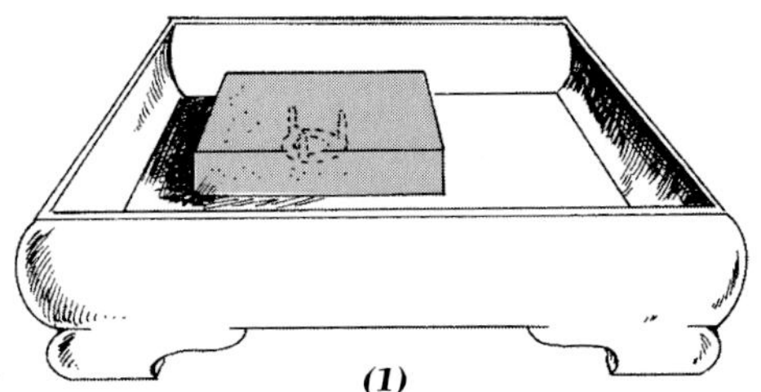

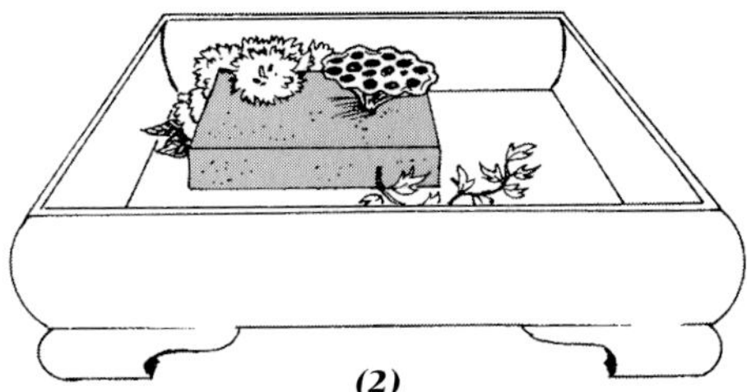

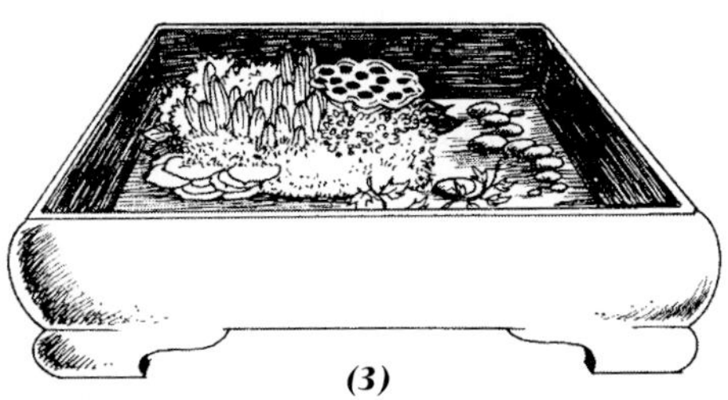

3-16 *Sheltered design—steps of construction: Step 1. Select a container that will allow materials to be arranged within the rim. Secure a small piece of floral foam. Step 2. Begin by grouping similar flowers and leaves. Cut stems short, allowing the heads to rest on the top of the foam. Use a variety of basing techniques. Step 3. Complete the design by concealing any exposed floral foam. Add stones, mosses, sand, twigs, and other materials.*

Sheltering is a technique that may be incorporated into many design styles. As shown in Figure 3-17, this contemporary technique can be used to create a protected, covered feeling. Bear grass, raffia, curly willow, and other linear materials lend themselves to simple sheltering techniques. Although covering or hiding part of a floral design is a strange concept, it is the covering that actually invites the viewer to discover what is underneath the shelter. The covering often suggests intimacy, mystery, or discernment and creates visual drama.

Pavé Design

The word *pavé* refers to a setting of jewelry in which gems are placed closely together so that little or no metal is visible (see Figure 3-18). Borrowed from jewelry making, a pavé floral design and the pavé technique refer to flowers, leaves, and other materials arranged closely together in a flat, jewel-like pattern so that no floral foam is visible. The tight clustering emphasizes contrasts in colors and textures.

Step 1 It will be helpful for you to plan out your pavé design pattern on paper first. Select a flat or low container. Next, cut a floral foam block in pieces to fit inside the container, as shown in Figure 3-19. Secure the foam into the container with anchor pins, waterproof tape, or pan glue. (Note: Often the pieces of floral foam will fit snugly into the container and remain secure enough without additional aids.)

(*Continued*)

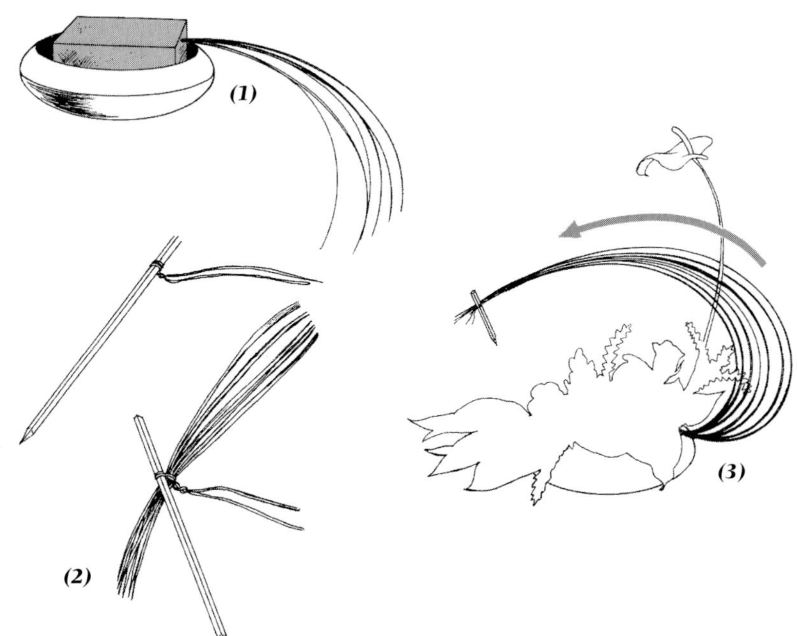

3-17 *Sheltering may be incorporated into a design using bear grass. Step 1. Insert a grouping of bear grass into one side of the floral foam. Step 2. The tips of the bear grass are gathered together using a wired wooden pick, as shown. Step 3. After completing the floral bouquet, insert the wooden pick with the gathered bear grass into the opposite side of the design.*

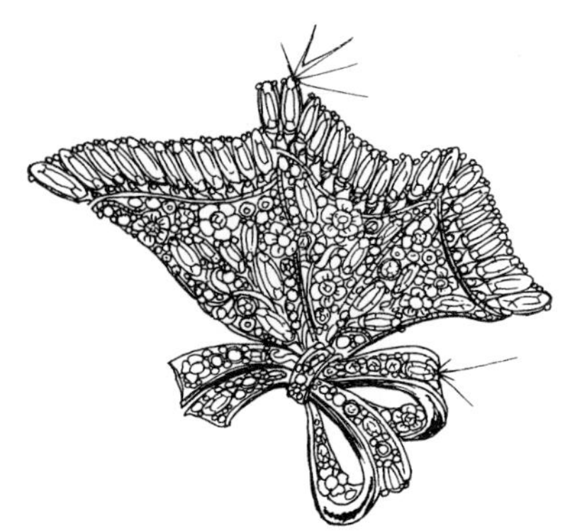

3-18 Borrowed from jewelry making, pavé refers to gems placed close together so that little or no metal is visible.

Step 2 Select flower heads, leaves, stones, mosses, and other materials that will create the desired pattern. Cut flower stems to a length of about one inch. Insert stems down into the foam. Flower heads should appear to be resting on the top of the foam. Similar materials should fit snugly against one another with no floral foam showing through.

Step 3 Continue adding flowers, leaves, stones, and other materials in the desired pattern. Contrast flower types, colors, and textures for added visual interest. The various sections should fit against one another, with no floral foam showing.

New Wave

New wave refers to any various new or experimental trends or movements. In floral design, new wave is a style of design featuring materials that have been changed with paint and glue, and altered and manipulated in other ways (often folded, bent, braided, curled, etc.). Flowers, foliage, and various materials are presented in unusual and bizarre configurations. Discordant and conflicting lines, colors, and geometric shapes are blended. Accessories and containers are generally peculiar in themselves, which adds to the eccentricity of the entire design. No rules exist for balance and proportion. Textures and patterns are often overemphasized. Ordinary materials are used in unexpected ways, adding to the visual drama.

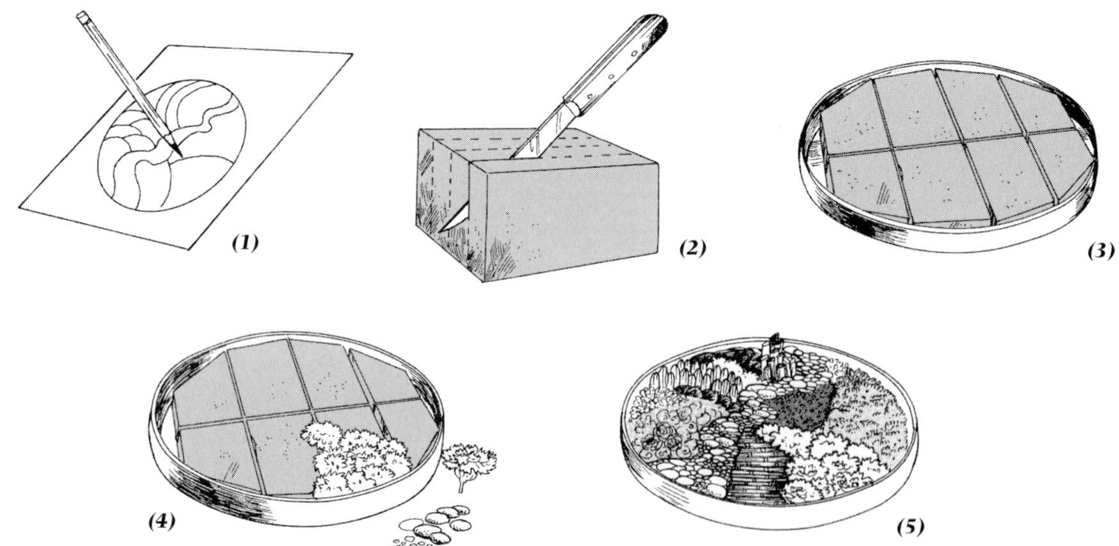

3-19 Pavé design—steps of construction: Step 1. Often it is helpful to plan the arrangement on paper first. Step 2. Cut long, low sections of floral foam. Step 3. Place the sections of floral foam into a low container, wedge tightly together so no further mechanics will be necessary. Step 4. Select short stemmed flowers, leaves, stones, mosses, and other materials that will create the desired pattern. Step 5. Continue adding materials in groups, concealing the foam. Contrast flower types, colors, and textures for added visual interest.

3-20 Basing is the process of placing materials at the foundation area of a design. The base of the design is generally contrasted with tall groups of materials.

Abstract Design

Often termed *free-form design*, an *abstract design* is a nonrealistic floral presentation that emphasizes shape, color, and texture. Nonfloral materials, such as metals, wires, plastics, glass, and mirrors, are frequently used to emphasize geometric forms. Often stems are crossed, petals or leaves may be stripped from stems, or materials are presented upside down. Abstract designs often become *interpretive designs*, reflecting the feelings and ideas of the designer.

Advanced Design Techniques

There are a number of *techniques* associated with contemporary advanced floral design. It is essential to know the names of these techniques or methods of construction and how they are carried out in order to make successful, beautiful, and distinctive designs.

Basing

Basing is the process of placing materials at the foundation area of a design. Many different techniques, such as clustering, pillowing, layering, and terracing, may be used to conceal the floral foam at the base of a design (see Figure 3-20). Basing provides visual stability and balance at the point from which taller stems emerge. The contrasting colors, textures, and shapes provide maximum visual appeal.

Juxtaposing Materials

For greater visual impact in color, form, and texture, similar materials may be *juxtaposed*. Several techniques are used in contemporary and advanced designs. Many of these methods, such as pillowing and layering, create an attraction area low at the base of an arrangement, while other techniques, such as grouping and zoning, stimulate visual interest at higher levels in a design.

Clustering, pillowing, tufting, and the pavé method share similar application methods; however, there are slight differences. *Clustering* is the process of placing materials closely together, usually at the design's base. However, materials may be clustered tightly together higher in the design, as shown in Figure 3-21. Clustering causes materials to lose their individual identity and function as a mass grouping of texture and color.

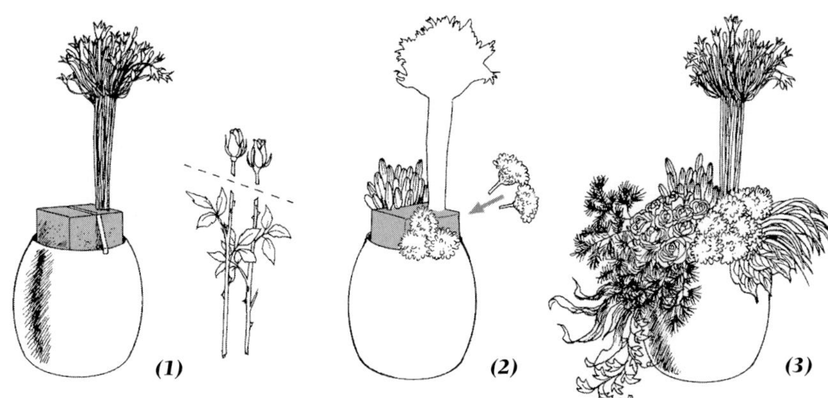

3-21 Clustering and a clustered design—steps of construction: Step 1. Choose an appropriate container for your design. Extend the floral foam above the container rim. Materials are placed close together in groups. Some stems may be tall, while other stems are cut short. Step 2. Arrange flowers in groups to increase visual contrast of color, texture, and form as materials may lose their identity and function as a mass grouping of texture and color. Step 3. Complete the design, concealing floral foam with clustered groups, both high and low in the design.

3-22 *Pillowing and a pillowed design—steps of construction: Step 1. Secure floral foam in a container. Insert a few tall stems to add interest and to contrast with the base of the design. Step 2. Insert materials close together in rounded groups or pillows. Step 3. Complete the design and conceal foam by snuggling various additional rounded groups of flowers, leaves, and other materials together.*

Pillowing is a specialized form of clustering and a popular basing technique. The clusters of flowers at the base of a design are arranged to form rounded hills or pillows. These rounded bunches of flowers form an unusual surface (see Figure 3-22). Individual pillow sections are comprised of only one type of material. When each pillow is a different color and texture to the adjacent pillows, color and texture are boldly emphasized.

Tufting is also a type of clustering, using bunches of short flower stems in a design to create a tufted, airy look. The material in each grouping is generally arranged in a radiating pattern. A tufted design is made more interesting with the addition of a few taller flowers, foliage stems, or curly branches rising from the arrangement (see Figure 3-23).

The *pavé* basing technique is a tight clustering method in which the surface of the bunches remains flat, rather than rounded as with pillowing and tufting (see Figure 3-24). Flowers, leaves, and other materials are clustered snugly against one another, creating a flat cobblestone effect.

Several basing techniques use similar materials arranged on top of one another, rather than side by side. As shown in Figure 3-25, similar materials are placed snugly on top of one another, with little space between individual leaves and flowers. *Terracing* is the placement of the same materials on top of one another, divided by space, giving a stair-step appearance. In contrast, *layering* is compactly overlapping the same materials (usually leaves) in a scalelike pattern. *Stacking* is placing one material horizontally on top of another to form piles or stacks.

Grouping and *zoning* are methods of bunching similar type materials together to emphasize forms and colors. In contrast to clustering techniques in which flowers lose their identity, grouping and zoning allow flowers continued visibility because they are arranged in a looser, less compact bunch. As shown in Figure 3-26, *grouping* or gathering similar flower and foliage types closely together gives individual forms and colors stronger emphasis. Generally, taller groups of materials are separated from one another with open or negative space.

Although similar to grouping, the technique of *zoning* places loose groups of materials in areas or zones within a composition. The quantity of material forming each group is generally restricted to allow a less compact bunch. Each group is slightly isolated from another through the use of negative space. Individual shapes and colors stand out with unusual independence (see Figure 3-27).

3-23 *A tufted design is made more interesting with the addition of a few taller flowers.*

3-24 *This elegant tufted pavé tray of flowers and foliage uses flat, parallel groupings and lines to create visual dynamics. From* Design with Flowers *© 1988 Herbert E. Mitchell; Photography: deGennaro Associates, Los Angeles, CA.*

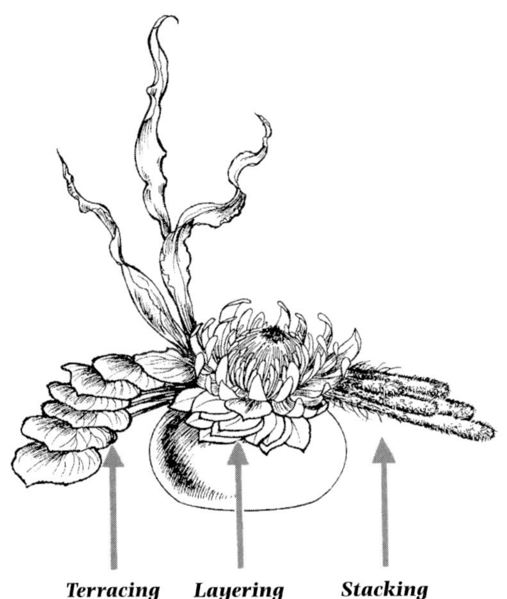

3-25 *Terracing, layering, and stacking techniques create an unusual pattern to increase visual movement in a design.*

3-26 *Grouping, a technique used in many design styles, emphasizes shapes, colors, and textures.*

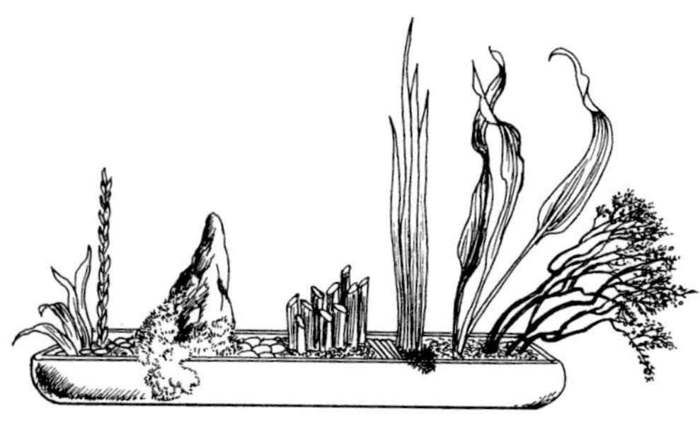

3-27 *Zoning provides for individual shapes and colors to stand out with unusual independence.*

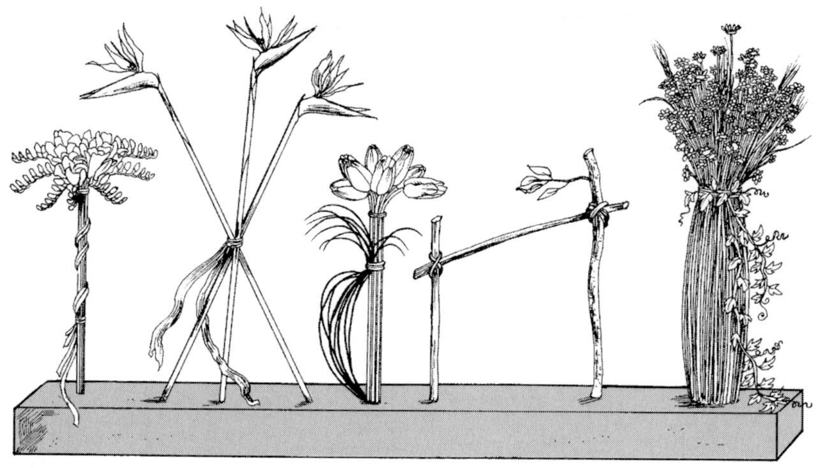

3-28 *Tying techniques serve a variety of ornamental and functional purposes in advanced contemporary designs.*

Uniting Materials

Several uniting or tying techniques, such as *banding, binding, bundling,* and *wrapping* may be used to unite or join materials (see Figure 3-28). Although the terms become confusing, these techniques all serve to combine materials. *Banding* is a method of tying materials together to draw attention to a certain area or element and is often merely decorative. Banding provides increased ornamentation and generally does not serve the purpose of physically joining materials together. *Binding,* in slight contrast to other tying techniques, is physically joining or fastening stems together. Although this method is functional and serves an actual purpose of holding stems together and in place, the binding of stems with *raffia,* ribbon, bear grass, and other materials is an attractive addition to floral arrangements. The *binding point* is the point or area where all stems come together or intersect, as in a *hand-tied bouquet*.

A simple analogy of banding and binding is that of wearing a bracelet compared to wearing a watch. Banding may be likened with wearing a decorative bracelet. It attracts the eye and is beautiful, but does not serve a functional purpose. In contrast, binding much like wearing a watch, serves a functional purpose yet may also be decorative.

Bundling is tying or wrapping similar materials together into one unit (such as wheat into sheaves or single cinnamon sticks into a larger group of several) and placing the bundled materials into a floral design.

Wrapping is a technique in which fabric, ribbon, raffia, metallic cord, and other materials are used to cover, coil, or twine a single stem or group of materials to achieve a decorative effect.

Strengthening Visual Movement

Several advanced techniques, such as *framing, shadowing,* and *sequencing,* may be used to increase rhythm and stimulate visual movement. *Framing* an arrangement is a technique in which material is placed in the perimeter of a design (see Figure 3-29). Although the flowers or branches that outline and frame a design initially lead the eye away from the focal area, the viewer is drawn back to the enclosed space. Framing enhances and calls attention to the materials in the central portion of the arrangement.

3-29 *Framing an arrangement is a technique in which material is placed in the perimeter of a design. Framed designs provide visual emphasis to the enclosed central material.*

Shadowing is a method of repetition. Also sometimes referred to as *mirroring*, this technique increases visual depth. Identical materials are placed closely behind and below the taller, front flower or leaf, forming a shadow (see Figure 3-30). Shadowing strengthens and draws attention to individual forms, colors, and textures.

When the materials in a floral design move in a progressing pattern of change, often referred to as *sequencing*, the eye is gradually led from one area to another. Sequencing may be easily accomplished through the gradation or transition of materials, especially color (from light to dark, or from one hue to another), size (from large to small), and height (from low to high). A gradual change in elements provides visual flow, increasing visual interest, and drama.

Because contemporary floral arrangement styles are closely connected with the styles of interior design, fashion design, and other types of design, it is vital for the floral artist to be aware of changing trends. Whether the trend in floral design is boldly tropical and exotic, quietly vegetative and natural, linear and high style, or peacefully sheltered, floral bouquets are indeed an art form that can express the times and feelings of a generation of people. Contemporary designs are, in essence, modern period-style floral arrangements.

Because contemporary designs are often unfamiliar and complex, specialized skills, mechanics, and techniques are often required (see Figure 3-31). Advanced, contemporary floral design provides a creative challenge to floral artists and requires designers to keep current in their knowledge of design styles and techniques.

3-30 *Shadowing is a method of repetition that promotes visual interest and movement. Identical materials are placed closely behind and below taller front materials.*

3-31 *Skill, knowledge, and experience are essential when constructing a floral design with expensive tropical flowers and foliage.*

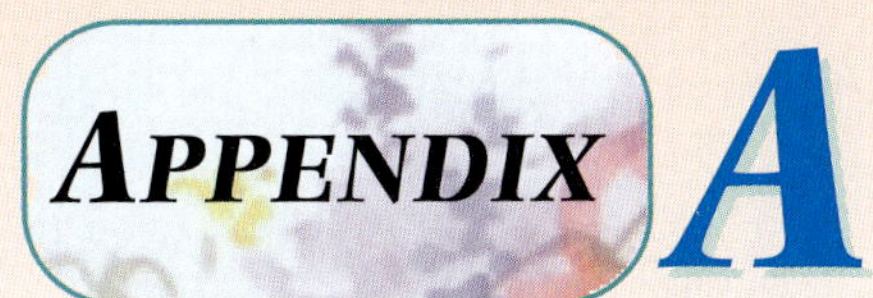

Appendix A Flowers

The aim of the flower appendix is to provide a wealth of information to help you become more familiar with and confident in pronouncing, selecting, and caring for cut flowers. Generic names and common names are listed alphabetically. The name of each ***genus*** is given first, (which is also the ***scientific name, botanical name,*** and the ***Latin name***) and is followed by a suggested pronunciation. The family in which it is placed, the derivation of the name, ***species,*** and ***common names*** are listed for each flower. Although most flowers are available year-round, some are limited to certain seasons, or have peak supplies in only one or two months of the year. A brief description of each flower is given along with approximate vase life and specific care and handling procedures. Common names are cross-referenced to the correct scientific name. The color plates, line illustrations, and photographs help ease identification.

Vase life can be dramatically increased by following several simple care and handling techniques. Choose flower varieties carefully. Often one variety, because of its genetic makeup will outlast another for several days or weeks. It is important to process flowers immediately upon arrival. Process the most expensive and wilt-sensitive flowers first. Studies show that cutting flowers under water and treating with a hydration solution, such as dipping newly cut stems for one second in Floralife Quick Dip® or for 30 to 60 minutes in Hydraflor®/100, will greatly speed the hydration and longevity of cut flowers. Place flowers in a warm preservative solution and let them stand outside the cooler for one to two hours. Next, place flowers in a 32–38°F cooler that has 80–90% relative humidity. Protect flowers from ethylene sources. Generally the grower, wholesaler, or retailer will treat flowers with an ethlyene inhibitor. For an in-depth look at care and handling, see Chapter 10.

A

ACACIA (see Plate 10)

(a-KAY-sha)
Family: Leguminosae or Fabaceae (pea family)
Name Origin: Named from the Greek word *akis* (sharp point), referring to the thorns.
Species: *dealbata, longifolia;* others
Common Names: acacia, mimosa, wattle
Availability: October through March
Description: Trees with clusters of fragrant, ball-shaped yellow flowers with gray-green, finely cut leaves. Branches of fluffy clusters work well as fillers; recut under water; place stems in hydration solution; cut flowers dry out and lose their fuzzy appearance as they age.
Vase Life: 4 to 5 days while in cooler, cover with plastic to retain moisture.

ACHILLEA

(ah-kil-LEE-ah)
Family: Compositae or Asteraceae (sunflower family)
Name Origin: Named after Achilles of Greek mythology, who is said to have used it medicinally.
Species: *filipendulina, millefolium;* others and hybrids
Common Names: yarrow, milfoil
Availability: July through September
Description: Flat-headed, yellow corymb flowers over a feathery foliage. Also available in pink, red, and white.
Vase Life: Long lasting, 7 to 10 days

ACONITUM

(ak-ah-NEE-tum)
Family: Ranunculaceae (buttercup family)
Species: *napellus;* others
Common Names: monkshood, aconite

ACHILLEA
yarrow

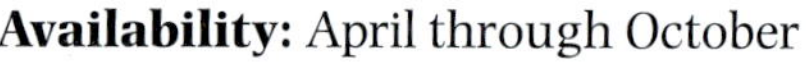

ACONITUM
monkshood

Availability: April through October
Description: Tall spikelike racemes of deep blue, hooded (helmet-shaped) florets. Also available in light blue, white, and cream varieties.
Vase Life: 5 to 10 days. Monkshood is extremely toxic. Wash hands after handling to get rid of toxic substances.

AFRICAN CORN LILY see ***Ixia***
AFRICAN DAISY see ***Gerbera***
AFRICAN LILY see ***Agapanthus***

AGAPANTHUS (see Plate 1)
(ag-a-PAN-thus)
Family: Amaryllidaceae (daffodil family)
Name Origin: From the Greek words *agape* (love) and *anthos* (flower)
Species: *africanus (A. umbellatus), orientalis;* others and hybrids
Common Names: lily of the Nile, African lily
Availability: March through August
Description: Large umbels of blue, funnel-shaped flowers in various tints. Also available in white. The stem of the agapanthus adds a strong line element with its bold-shaped yet airy flower head. Mixes well with other flowers in many different design styles.
Vase Life: 4 to 7 days, with individual florets blooming continuously over several days; ethylene sensitive.

ALCHEMILLA
(al-ke-MIL-la)
Family: Rosaceae (rose family)
Species: *mollis;* others
Common Name: lady's mantle
Availability: Spring to early fall with peak supplies May through July
Description: Feathery sprays (compound cymes) of small, yellowish or greenish flowers. Useful as an airy filler.
Vase Life: Long lasting, 7 to 10 days

ALLIUM (see Plate 1)
(AL-ee-um)
Family: Liliaceae (lily family)
Name Origin: From the Greek name for garlic
Species: *giganteum, sphaerocephalon;* others
Common Names: allium, onion flower, garlic, flowering onion
Availability: April through September
Description: Bulbs producing globe-shaped umbels of purple and pink flowers on leafless stems. Also available in blue, white, and yellow varieties. Most smell like onion when cut, but odor generally dissipates.
Vase Life: 5 to 10 days

ALSTROEMERIA
Peruvian lily
ANEMONE
windflower
ALLIUM
AGAPANTHUS
lily of the Nile

ALCHEMILLA
lady's mantle

ALPINIA (see Plate 2)

(al-PIN-ee-ah)
Family: Zingiberaceae (ginger, cardamom, and turmeric)
Species: *purpurata, zerumbet;* others
Common Names: ginger, ostrich plume, torch ginger, shell ginger
Availability: Year-round
Description: The flower head consists of shiny red or pink bracts at the end of a heavy, thick stem, often up to 36 inches long. Ginger provides a strong vertical line. The colored bracts provide emphasis. If flowers appear wilted, immerse the entire stem and bracts in room-temperature water for 15 to 30 minutes.
Vase Life: Incredibly long lasting, up to 3 weeks

ALSTROEMERIA (see Plate 1)

(al-stre-MEAR-ee-ah)
Family: Alstroemeriaceae (alstroemeria family)
Name Origin: Named after Baron Claus Alstroemer (1736–1794).
Species: *heamantha, aurantiaca, pelegrina;* others and hybrids
Common Names: alstroemeria, alstro, Peruvian lily, Inca lily, lily of the Incas
Availability: Year-round
Description: Umbelled clusters of delicate, trumpet-shaped flowers borne at the end of short flower stalks that spray off of a single stem. Available in many colors with many intermediate colors. Most varieties are freckled or streaked with contrasting colors.
Vase Life: Long lasting, up to 2 weeks, with individual flowers lasting 5 to 7 days each. Ethylene sensitive, with leaves often yellowing prematurely; remove lower leaves.

AMARANTHUS

(am-a-RAN-thus)
Family: Amaranthaceae (cockscombs and celosias)
Name Origin: From the Greek word *amarantos* (unfading), referring to the long-lasting flowers.
Species: *hypochondriacus, tricolor, caudatus;* others and hybrids
Common Names: amaranth; prince's feather *(A. hypochondriacus);* Joseph's coat (*A. tricolor);* love lies bleeding, cat's tail, tassel flower *(A. caudatus)*
Availability: Summer and autumn
Description: Erect and brushlike flower racemes and spikes up to 6 inches long. Available in red, green, and cream. The species *A. caudatus* and hybrids are slender, drooping red-flower racemes up to 16 inches long.
Vase Life: Long lasting, 7 to 10 days

AMARANTHUS
amaranth

ANTHURIUM

STRELITZIA
bird of paradise

ALPINIA
ginger

AMARYLLIS (see Plate 3)

(am-a-RIL-is)
Family: Amaryllidaceae (daffodil family)
Name Origin: Named after a shepherdess in Greek mythology.
Species: *belladonna;* see *Hippeastrum* for the larger trumpet-shaped flower that is commonly called *Amaryllis.*
Common Names: belladonna lily, naked lady lily, cape belladonna
Availability: May through November
Description: Trumpet-shaped flowers, each 2 to 3 inches across on a single thick, leafless stem. Available in pink, red, and white. Belladonna lilies are fragrant and attractive in many design styles, especially parallel and vegetative designs where their forms are apparent.
Vase Life: Long lasting, up to 10 days

AMAZON LILY see *Eucharis*

AMMI (see Plate 21)

(AM-me)
Family: Umbelliferae or Apiaceae (carrot family)
Species: *majus;* (similar are the species *Daucus carota* and *D. sativus)*
Common Names: Queen Anne's lace, bishop's weed; wild Queen Anne's lace *(Daucus)*
Availability: Year-round
Description: Delicate, white, compound umbels, 3 to 6 inches across. Queen Anne's lace is more airy than *Daucus* varieties. Both *Ammi* and *Daucus* work well as fillers.
Vase Life: 3 to 5 days

ANANAS (see Plate 21)

(a-NA-nas)
Family: Bromeliaceae (pineapple family)
Species: *nanus, bracteatus stivus, comosus;* others
Common Names: ornamental pineapple, dwarf pineapple
Availability: Year-round
Description: Large or small pink, white, or variegated head resembles the edible pineapple. Provides an immediate focal point through shape and texture. Use caution—plants have thorny spikes.
Vase Life: Long lasting, 1 to 3 weeks

ANEMONE (see Plate 1)

(a-NEM-oh-nee)
Family: Ranunculaceae (buttercup family)
Name Origin: Named after Adonis, also called Naamen, a handsome young man of Greek mythology who was loved by Aphrodite. He was killed by a wild boar and his blood is said to have given rise to the blood-red flowers.
Species: *coronaria;* others and hybrids
Common Names: windflower, lily of the field, poppy anemone
Availability: October through May
Description: Cup-shaped solitary flowers open up flat. Petals surround a dark center. Available in vibrant colors of red, pink, blue, purple, as well as white. Many anemones curve toward the light.
Vase Life: 3 to 7 days. Keep cool to prevent premature wilting and drooping.

ANETHUM

(a-NAY-thum)
Family: Umbelliferae or Apiaceae (carrot family)
Species: *graveolens*
Common Name: dill
Availability: Year-round
Description: Compound umbel clusters of yellow flowers, similar to Queen Anne's lace, on stems up to 36 inches tall. A useful, long-lasting filler.
Vase Life: 7 to 10 days

ANIGOZANTHOS (see Plate 14)

(a-nee-go-ZAN-thus)
Family: Haemodoraceae (kangaroo paw family)
Name Origin: From the Greek words *anoigo* (to open) and *anthos* (flower), referring to the flowers that are widely open.
Species: *flavidus, pulcherrimus, rufus;* others and cultivars
Common Name: kangaroo paw
Availability: Year-round
Description: Unusual red, purple, green, or yellowish, fuzzy flowers borne in one-sided racemes and spikes. Each blossom has a covering of short, colored fur providing a delicate texture contrast. Combines easily with other exotic flowers.
Vase Life: Long lasting, up to 2 weeks

ANNUAL DELPHINIUM see *Consolida*

ANTHURIUM (see Plate 2)

(an-THUR-ee-um)
Family: Araceae (the aroids)
Name Origin: From the Greek words *anthos* (flower) and *oura* (tail), referring to the tail-like inflorescence or spadix.
Species: *andraeanum, scherzeranum*
Common Names: anthurium, tail flower, flamingo flower, painted tongue, painter's palette, peace lily
Availability: Year-round
Description: Anthurium flowers consist of a modified, shiny, colorful leaf called a "spathe." The spathe is usually heart-shaped or arrow-shaped, in reds, pinks, white, bicolors, and tricolors. The true flowers are on a cylindrical, long *spadix* that is usually yellow, but sometimes appears in other colors. The size of the spathe varies from $2\frac{3}{4}$ to 6 inches or

AMARYLLIS
belladonna lily
CALLISTEPHUS
China aster
ASTER
monte cassino
ASTILBE
false spirea

more. These flowers lend a dramatic focal point and look best when used alone or with other exotic flowers.
Vase Life: Anthuriums are extremely hardy cut flowers, lasting 2 to 3 weeks or more; however, care must be taken in handling anthuriums as they bruise easily. Anthuriums are accustomed to humid and warm conditions. Mist often with water and do not store below 45°F. If bracts look wilted, immerse entire stem in water for 15 to 30 minutes.

ANTIRRHINUM (see Plate 24)

(an-tee-RYE-num)
Family: Scrophulariaceae (foxglove family)
Name Origin: From the Greek words *anti* (like) and *rhis* (snout), referring to the appearance of the individual florets.
Species: *majus;* cultivars
Common Names: snapdragon, snaps
Availability: Year-round
Description: A spiked, terminal raceme with florets 1½ inches long that are tubular with rounded upper and lower lips (when squeezed together they snap open and shut). Available in many colors. Snapdragons are geotropic (bend upright away from gravity); store or arrange in a vertical position to prevent curvature. Removing the very top bud can help deter bending.
Vase Life: Capable of a long vase-life of up to 2 weeks; ethylene sensitive.

AQUILEGIA

(ak-wi-LEE-ji-a)
Family: Ranunculaceae (buttercup family)
Name Origin: From the Latin word *aquila* (eagle). The flower petals resemble the claws of eagles and other birds of prey.
Species: *caerula, canadensis, chrysantha, flabellata;* others and hybrids
Common Name: columbine
Availability: April through September
Description: Terminal, bonnet-shaped flowers consisting of five petals, each with its own protruding spur. The flowers are 1 to 3 inches long. Available in white, pink, yellow, blue, and purple.
Vase Life: 3 to 5 days; do not store in cooler.

AQUILEGIA
columbine

ARUM LILY see ***Zantedeschia***

ASCLEPIAS

(as-KLEE-pee-us)
Family: Asclepiadacea (milkweeds and wax plant)
Name Origin: From Asklepios, the Greek god of medicine, referring to medicinal properties.
Species: *tuberosa;* others
Common Names: milkweed, blood flower, silkweed
Availability: Year-round, with peak supplies June through September
Description: Tiny, waxy, orange flowers appear in rounded, umbellate cyme clusters. Adds mass and accent to designs; can also be used as a filler flower.
Vase Life: 3 to 7 days; florets continue to open; remove old or dying florets. Stems exude a milky sap (hence the name milkweed); dip stems in hot water and place in warm perservative solution.

ASIAN LILY see ***Lilium***

ASTER (see Plate 3)

(AS-ter)
Family: Compositae or Asteraceae (sunflower family)
Name Origin: From the Latin word *aster* (star), referring to the flowers.
Species: *cordifolius, ericoides, novi-belgii;* others and hybrids
Common Names: monte cassino aster (*A. ericoides* 'Monte Cassino'); Michaelmas daisy (*A. novi-belgii*)
Availability: July through December
Description: Numerous daisylike flowers ¾ to 1 inch across in double or single types. Available in a wide variety of colors, usually with a yellow center. The flowers are clustered racemes, corymbs, or panicles.
Vase Life: Long lasting, up to 10 days

MOLUCCELLA
bells of Ireland
BOUVARDIA
RUDBECKIA
black-eyed Susan
TRITELEIA
brodiaea

ASTILBE (see Plate 3)

(a-STIL-bee)
Family: Saxifragaceae (currants, hydrangeas, and saxifrages)
Name Origin: From the Greek words *a* (without) and *stilbe* (brillance), meaning the individual flowers are extremely small.
Species: *chinensis;* others and hybrids
Common Names: false spirea, goat's beard, meadowsweet
Availability: Year-round
Description: Pyramidal feather plumes. Flowers are borne in loose, pyramidal panicles on slender stems. Available in white, pink, and red. Excellent filler flower.
Vase Life: 5 to 7 days

ASTRANTIA

(a-STRAN-tee-ah)
Family: Umbelliferae or Apiaceae (carrot family)
Name Origin: From the Latin word *aster* (star) referring to the starlike flowers.
Species: *major;* others and hybrids
Common Name: masterwort
Availability: April to September
Description: Delicate, star-shaped flowers in clustered umbels. Available in white, pinks, and reds. Useful as a mass or filler flower. Up close, Astrantia has a slight, sometimes unpleasant, odor.
Vase Life: 4 to 7 days

ATRIPLEX

(AH-tri-plex)
Family: Chenopodiaceae (sugar beet family)
Name Origin: Greek name for *A. hortensis* herbs and shrubs.
Species: *lumex;* others
Common Names: saltbush, greasewood
Availability: November through April
Description: Tall erect stems with open loose flower heads; available in red, green, and gray; sometimes used as cut foliage.
Vase Life: 5 to 7 days

ATRIPLEX saltbush

B

BABY'S BREATH see *Gypsophila*

BACHELOR'S BUTTON see *Centaurea*

BANKSIA (see Plate 20)

(BANK-see-ah)
Family: Proteaceae (proteas, banksias, and grevilleas)
Name Origin: Named after botanist Sir Joseph Banks (1743–1820).
Species: *ashbyi, baxteri, coccinea, collina, ericifolia, menziesii, speciosa;* many others
Common Names: protea, bird's nest, giant bottlebrush
Availability: Year-round
Description: Conelike, dense terminal spikes in reds, oranges, and yellows. Provides an immediate focal point. Foliage has textural appeal.
Vase Life: Extremely long lasting, 2 to 4 weeks

BARBERTON DAISY see *Gerbera*
BELLADONNA LILY see *Amaryllis*
BELLFLOWER see *Campanula*
BELLS OF IRELAND see *Moluccella*
BILLY BUTTONS see *Craspedia*
BIRD OF PARADISE see *Strelitzia*
BIRD'S NEST see *Banksia*
BLACK-EYED SUSAN see *Rudbeckia*
BLANKET FLOWER see *Gaillardia*
BLAZING STAR see *Liatris*
BLOOD FLOWER see *Asclepias*
BLUEBELL see *Scilla*
BLUE LACEFLOWER see *Trachymene*
BLUE THROATWORT see *Trachelium*

BORONIA

(ba-RONE-knee-ah)
Family: Rutaceae (citrus fruit family)
Species: *heterophylla*
Common Name: boronia
Availability: January to September, with peak supplies in the spring
Description: Long spikes of bright hot pink, light pink, or purple bell-shaped flowers nestled up and down the stem of soft, needlelike leaves; similar to heather but softer with larger flowers; citrusy scent.
Vase Life: 1 to 2 weeks; susceptible to dehydration; condition in hydrating solution outside the cooler; air dries nicely.

BORONIA

CIRSIUM
plume
thistle
CAMPANULA
chimney bells
CAMELLIA
IBERIS
candytuft
ZANTEDESCHIA
calla lily

BOUVARDIA (see Plate 4)

(boo-VAR-dee-ah)
Family: Rubiaceae (gardenias, coffee, and quinine)
Name Origin: Named after Dr. Charles Bouvard (1572–1658).
Species: *longiflora* hybrids
Common Name: bouvardia
Availability: Year-round
Description: Small, tubular flowers with spreading starlike petals in terminal cyme clusters. Generally fragrant. Available in white, pinks, oranges, and reds. Excellent filler flower. Often used in wedding and corsage work.
Vase Life: 7 to 10 days; sensitive to water stress and cool temperatures; ethylene sensitive; prefers preservative solution to the use of floral foam. Remove foliage on lower stem.

BRAIN FLOWER see *Celosia*
BRODIAEA see *Triteleia*
BROOM see *Cytisus*

BUPLEURUM

(boo-PLUR-rum)
Family: Umbelliferae or Apiaceae (carrot family)
Name Origin: From the Greek word *boupleuros* (ox rib).
Species: *rotundifolium, griffithii;* others
Common Name: thoroughwax
Availability: Year-round
Description: Multibranched green filler flower and foliage; tiny yellow crown of flowers at the ends of green, leafy stems; adds a light, airy, informal look.
Vase Life: 7 to 10 days; wash lower stems and recut under water; immediately place in hydration solution; condition outside of cooler for 1 to 2 hours.

BUPLEURUM
thoroughwax

BUTTERFLY ORCHID see *Oncidium*
BUTTON MUM see *Chrysanthemum*
BUTTON SNAKEROOT see *Liatris*

C

CALENDULA

(ka-LEN-dew-la)
Family: Compositae or Asteraceae (sunflower family)
Name Origin: From the Latin word *calandae* (first day of the month), referring to its long flowering period.

CALENDULA
pot marigold

Species: *officinalis;* many hybrids
Common Name: pot marigold
Availability: April through November
Description: Daisylike double-petaled head flowers. Available in bright yellows and oranges.
Vase Life: 5 to 7 days

CALLA LILY see *Zantedeschia*

CALLISTEPHUS (see Plate 3)

(ka-LIS-te-fus)
Family: Compositae or Asteraceae (sunflower family)
Name Origin: From the Greek words *kallos* (beautiful) and *stephanus* (crown), referring to the showy solitary flower heads.
Species: *chinensis* hybrids
Common Names: China aster, aster, Matsumoto aster
Availability: June through September
Description: Large, solitary flower heads. Available in a variety of shapes and colors.
Vase Life: 5 to 10 days; remove any excess foliage for longer vase life.

CALLUNA see *Erica*
CALYCINA see *Thryptomene*

CAMELLIA (see Plate 5)

(ka-MEL-ee-a)
Family: Theaceae (tea, camellias, and franklinia)
Name Origin: Named after pharmacist George Joseph Kamel (1661–1706), who studied the Philippines flora.

CHRYSANTHEMUM—Some of the most common inflorescence forms grown commercially.

Species: *japonica, reticulata, sinensis;* others
Common Name: camellia
Availability: January and February
Description: Solitary waxy flowers. Available mostly in pinks, reds, and white. Single flowers are excellent for weddings and corsages. Larger flowering branches add interest to large designs.
Vase Life: Single flowers that are prepackaged, last only 1 to 2 days. However, woody branches with flower buds can last more than a week.

CAMPANULA (see Plate 5)

(kam-PAHN-ew-lah)
Family: Campanulaceae (bellflower family)
Name Origin: From the Latin word *campana* (bell), referring to the bell-shaped flowers.
Species: *glomerata, persicifolia, pyramidalis;* others
Common Names: bellflower, chimney bells, Canterbury bells
Availability: April through August
Description: Whether clustered or spikelike, these work well in mixed summer arrangements. *C. glomerata* has a bold clustered shape that works well in contemporary design styles.
Vase Life: 5 to 7 days

CANDYTUFT see *Iberis*

CAPE JASMINE see *Gardenia*

CAMPANULA GLOMERATA

CARNATION see *Dianthus*

CARTHAMUS (see Plate 24)

(CAR-tha-mus)
Family: Compositae or Asteraceae (sunflower family)
Species: *tinctorius*
Common Names: safflower, false saffron
Availability: June through November
Description: These thistlelike flowers are about 1 inch across with a green globular center, where thin orange or yellowish petals emerge. Safflowers add interesting texture to designs. Useful as an accent or filler.
Vase Life: Long lasting, 1 to 2 weeks. Removal of excess foliage increases vase life.

CASPIA see *Limonium*

CAT'S TAIL see *Amaranthus*

CATTLEYA (see Plate 19)

(KAT-lee-ah)
Family: Orchidaceae (orchid family)
Name Origin: Named after horticultural patron, William Cattley (early 1800s).
Species: *aurantiaca, bicolor, intermedia, labiata;* others and hybrids
Common Name: corsage orchid
Availability: Year-round
Description: The flowers measure 3 to 6 inches across. Available in white, lavender, pink, and yellow. Their shape is exotic and interesting, with showy sepals and petals. A broad ruffled lip is in the center. Generally used for corsage and wedding flowers but can be incorporated into arrangements. Generally packaged singly with individual water tubes.
Vase Life: Long lasting, 5 to 10 days with direct water supply

CELOSIA

(see-LO-si-ah *or* see-LO-shee-ah)
Family: Amaranthaceae (cockscombs and celosias)
Name Origin: From the Greek word *keleos* (burning), referring to the brightly colored flowers.
Species: *argentea, cristata, plumosa;* hybrids
Common Names: brain flower, cockscomb, woolflower
Availability: July through November
Description: Available in reds, yellows, and oranges. *C. cristata* with its crested shape adds interesting form and texture to contemporary designs. The feathery plumes of *C. plumosa* work well as fillers and accents.
Vase Life: 5 to 7 days

CROCOSMIA
montbretia
CENTAUREA
cornflower or
bachelor's button
NARCISSUS
paperwhites
NARCISSUS
daffodil
DELPHINIUM

CELOSIA PLUMOSA

CELOSIA CRISTATA

CENTAUREA (see Plate 7)

(sent-ta-REE-a)
Family: Compositae or Asteraceae (sunflower family)
Name Origin: From the Greek word *kentaur* for Centaur, who is said to have used it medicinally.
Species: *cyanus, macrocephala;* others
Common Names: cornflower, bachelor's button, bluebottle
Availability: February through September
Description: Small, thistlelike head flowers. Available in blue, pink, yellow, and white. Useful as a filler.
Vase Life: 5 to 7 days

CHAENOMELES (see Plate 22)

(kee-NAHM-el-eez)
Family: Rosaceae (rose family)
Name Origin: From the Greek words *chaina* (to gape) and *melon* (apple), referring to the belief the fruit was split.
Species: *cathayensis, japonica, speciosa*
Common Name: flowering quince
Availability: January to April
Description: Clusters of pink, red, or white flowers on woody branches. Works well in oriental designs.
Vase Life: 3 to 10 days

CHAMELAUCIUM (see Plate 28)

(cham-ee-LAW-si-um)
Family: Myrtaceae (myrtles, eucalyptus, and cloves)
Species: *uncinatum, ciliatum, pheliferum;* others
Common Name: waxflower
Availability: September through May
Description: Heathlike shrubs with star-shaped flowers in white, pinks, lavenders, and bicolors. Densely clustered along woody branching stems. Needlelike foliage. An excellent filler flower. Many have lemon scent.
Vase Life: 7 to 10 days; ethylene sensitive.

CHELONE

(kee-LO-nee *or* shel-LONE)
Family: Scrophulariaceae (foxglove family)
Name Origin: From the Greek word *chelone* (turtle), referring to the corolla that is shaped like a turtle's head.
Species: *obliqua*
Common Names: turtlehead, snakehead
Availability: July through October
Description: Pink flowers are similar in appearance to snapdragons; however, florets are clustered closer together in terminal, spikelike racemes.
Vase Life: 5 to 7 days

DIANTHUS—Some of the most common types grown commercially.

CHELONE
turtlehead

CHIMNEY BELLS see ***Campanula***
CHINA ASTER see ***Callistephus***
CHINCHERINCHEE see ***Ornithogalum***
CHRISTMAS ROSE see ***Helleborus***

CHRYSANTHEMUM (see Plate 6)

(kris-ANTH-e-mum)

Family: Compositae or Asteraceae (sunflower family)

Name Origin: From the Greek words *chrysos* (golden) and *anthos* (flower).

Species: *morifolium, frutescens, coccincum, parthenium, carinatum;* others and cultivars

MATRICARIA feverfew

Common Names: mum or florist's chrysanthemum *(C. morifolium)* includes spray varieties (duet or anemone, button, daisy, cushion [formerly called pompom], spider, and exotic varieties like starburst) as well as single-head varieties (Fuji, incurve or football mum, and mefo). Many other types include daisy, marguerite daisy *(C. frutescens)*; Shasta daisy *(C. maximum);* feverfew *(C. parthenium, Matricaria capansis)*, pyrethrum *(C. coccineum);* oxeye daisy *(Leucanthemum).*

Availability: Year-round

Description: Available in a wide variety of shapes, colors, and textures. Smaller spray types work well as fillers and accents. Larger solitary flowers provide mass and emphasis.

Vase Life: Long lasting, 1 to 3 weeks; Marguerite daisies are heavy drinkers. Remove wilted foliage from stems. To prevent shattering, spray the back and top of flower heads with an aerosol designed for this purpose. Do not pound or smash stems.

CIRSIUM (see Plate 5)

(SIR-cee-um)

Family: Compositae or Asteraceae (sunflower family)

Species: *japonicum;* others and hybrids

Common Names: plume thistle, plumed thistle. *Cnicus benedictus* is a similar thistlelike flower with spiny, leafy bracts.

Availability: May through November

Description: Small thistlelike flower heads. Available in pinks and purples. Useful as a filler. Provides interesting accent because of texture and shape.

Vase Life: Long lasting, 1 to 2 weeks

CLARKIA (see Plate 13)

(KLARK-ee-ah)

Family: Onagraceae (clarkias, fuchias, and evening primroses)

Name Origin: Named after Captain William Clark (1770–1838).

Species: *amoena, concinna;* others

Common Names: godetia, satin flower, farewell-to-spring. *Godetia* (goh-DEE-shee-ah) comprises a subgenus of *Clarkia* and is sometimes listed separately.

Availability: May through August

Description: Clustered funnel-shaped flowers about 2 inches across with a papery texture. Available in a wide range of colors and bicolors. Godetia provides emphasis or mass, and can also be used as a filler.

Vase Life: 5 to 7 days

CNICUS see ***Cirsium***
COBRA HEAD see ***Sarracenia***
COCKSCOMB see ***Celosia***
COFFEE BEAN BERRY see ***Hypericum***
COLEONEMA see ***Diosma***
COLUMBINE see ***Aquilegia***
CONE FLOWER see ***Rudbeckia***

DIGITALIS
foxglove

EREMURUS
foxtail lily

CYNARA
globe artichoke

CONSOLIDA (see Plate 16)

(kon-SO-li-da)
Family: Ranunculaceae (buttercup family)
Name Origin: From the Latin word *consolida* (to make whole), referring to its medicinal properties.
Species: *ambigua (Delphinium ajacis); orientalis (D. orientale); regalis (D. consolida). Consolida* florets and leaves differ slightly from *Delphinium;* the two upper petals are united into one, and they lack the lower two petals.
Common Name: larkspur
Availability: June through September
Description: Spikelike racemes in blues, lavenders, pinks, and white. Excellent for adding mass and line.
Vase-Life: 7 to 10 days; ethylene sensitive.

CONVALLARIA (see Plate 18)

(kahn-val-AIR-ee-ah)
Family: Liliaceae (lily family)
Name Origin: From the Latin word *convallis* (valley).
Species: *majalis*
Common Name: lily of the valley
Availability: Year-round
Description: Fragrant and delicate white or pink bell-shaped florets along short, terminal, one-sided racemes. Popular for wedding and corsage work. Also an excellent accent in contemporary or vegetative design styles.
Vase Life: 2 to 6 days

COSMOS

CORNFLOWER see *Centaurea*

CORN LILY see *Ixia*

CORSAGE ORCHID see *Cattleya*

COSMOS

(KAHZ-mohs)
Family: Compositae or Asteraceae (sunflower family)
Name Origin: From the Greek word *kosmos* (beautiful).
Species: *bipinnatus, sulphureus;* others
Common Name: cosmos
Availability: June through August
Description: Solitary flowers with a single row of petals in white, yellow, reds, and pinks surrounding a yellow center. Useful for adding mass; works well as a filler in summer bouquets.
Vase Life: 4 to 6 days

COSTUS

(KAWS-tus)
Family: Zingiberaceae (ginger family)
Species: *megalobractiatus, pulverulentus, spicatus;* others
Common Names: spiral ginger, kiss of death, spiral flag
Availability: Year-round
Description: Spirally twisted conelike bracts, available in red, orange, yellow, pink, and white. These unusual flowers provide emphasis.
Vase Life: 5 to 10 days or longer

CRANE LILY see *Strelitzia*

CRASPEDIA

(cras-PEH-dee-ah)
Family: Compositae or Asteraceae (sunflower family)
Species: *globosa*
Common Names: billy buttons, Australian bachelor buttons
Availability: August through October
Description: Long-lasting ball-shaped yellow flowers; can easily dry
Vase Life: 1 to 2 weeks

CRASPEDIA
billy buttons

ACACIA
EUPHORBIA
scarlet plume
EUCALYPTUS PODS
FORSYTHIA
golden bells

CROCOSMIA (see Plate 7)

(kro-KOS-mee-ah)
Family: Iridaceae (iris family)
Name Origin: From the Greek words *krokos* (saffron) and *osme* (smell). When dried, they smell of saffron.
Species: *masoniorum*
Common Names: montbretia, coppertip
Availability: June through November
Description: Trumpet-shaped flowers in one-sided, spike-like patterns. Available in shades of scarlet, orange, and red. Useful in many design styles as a filler, line, or accent.
Vase Life: 1 to 2 weeks

CURCUMA

(ker-CUE-mah)
Family: Zingiberaceae (ginger family)
Species: *alismatifolia, domestica;* others
Common Names: summer tulips, Thia tulips, Siam tulips, hidden lilies, queen lilies
Availability: April through September
Description: Light pink and other colored bracts; looks like ginger.
Vase Life: Long lasting, 1 to 3 weeks or more; do not place in the cooler.

CURCUMA
summer tulips

CYMBIDIUM (see Plate 19)

(sim-BID-ee-um)
Family: Orchidaceae (orchid family)
Name Origin: From the Greek word *kymbe* (boat), referring to the hollowed lip.
Species: hybrids
Common Name: cymbidium
Availability: Year-round
Description: Spray of orchid flowers. Usually sold singly with individual water tubes. Wide variety of colors. Useful for corsage and wedding flowers but can be incorporated into arrangements.
Vase Life: Extremely longlasting, 1 to 3 weeks with good water supply. Avoid temperatures below 38° F. Do not damage the stigma, which results in premature wilting.

CYNARA (see Plate 9)

(SIN-ah-rah)
Family: Compositae or Asteraceae (sunflower family)
Name Origin: From the Latin name for these perennial herbs.
Species: *scolymus*
Common Names: globe artichoke, artichoke, thistle
Availability: July through October

CYTISUS
genista

Description: Tight green, thistlelike spiney heads, 2 to 4 inches in diameter, from which purple and blue flowers emerge. These add interesting texture to designs. Useful as a focal point, accent, or filler.
Vase Life: 7 to 10 days

CYPRIPEDIUM see ***Paphiopedilum***

CYTISUS

(SIT-is-us)
Family: Leguminosae or Fabaceae (pea family)
Name Origin: From the Greek word *kytisos*, the name for these and similar shrubs.
Species: *canariensis (Genista canariensis);* for *C. scoparius (G. scoparia)* see in Appendix B under *Cytisus*.
Common Names: genista, broom, sweet broom
Availability: January through May
Description: Small, fragrant pea-shaped florets on wiry, leafless branches. Available in white and yellow, along with other dyed colors of pinks and purples. Useful as a filler and for soft curving lines in many design styles.
Vase Life: 1 to 2 weeks

D

DAFFODIL see ***Narcissus***

DAHLIA (see Plate 15)

(DAHL-yah)
Family: Compositae or Asteraceae (sunflower family)

GAILLARDIA
blanket flower
FREESIA
GARDENIA
GERBERA
African daisy

Name Origin: Named after Swedish botanist Dr. Anders Dahl (1751–1789).
Species: many hybrids and cultivars
Common Name: dahlia
Availability: July through November
Description: The wide variety of shapes, sizes, colors, and textures of these head flowers allow many design uses. Can add mass, provide emphasis, and accent. Smaller varieties work well as fillers.
Vase Life: Varies greatly, 2 to 10 days

DAISY see ***Chrysanthemum***
DANCING DOLL see ***Oncidium***
DAUCUS see ***Ammi***

DELPHINIUM (see Plate 7)
(del-FIN-ee-um)
Family: Ranunculaceae (buttercup family)
Name Origin: From the Greek word *delphis* (dolphin), referring to the shape of the flowers.
Species: *elatum, D. x belladonna, cardinale, grandiflorum;* others. For *D. ajacis*, see *Consolida ambigua* and *C. orientalis.*
Common Name: delphinium
Availability: Year-round
Description: Spikelike racemes available in white and various tints and shades of blue, lavender, and pink. Useful as a line flower or in adding mass. Some varieties work well as fillers. Many design uses for a variety of styles.
Vase Life: Varies greatly with species, from 3 to 4 days to 2 weeks; ethylene sensitive; florets can easily shatter, so handle with care.

DENDROBIUM (see Plate 19)
(den-DRO-be-um)
Family: Orchidaceae (orchid family)
Name Origin: From the Greek words *dendron* (tree) and *bios* (life), referring to their epiphytic growth (an epiphyte is a plant that grows on another plant but is not a parasite and produces its own food by photosynthesis).
Species: *bigibbum, nobile, phalaenopsis;* others and hybrids
Common Names: dendrobium, Singapore orchid
Availability: Year-round
Description: Spray orchids in lavenders, pinks, and white. Spray racemes provide slightly curving lines. Commonly used with exotic or tropical flowers. Individual blossoms are used for corsages, boutonnieres, and wedding designs.
Vase Life: 7 to 10 days

DESERT CANDLE see ***Eremurus***

DIANTHUS (see Plate 8)
(die-ANTH-us)
Family: Caryophyllaceae (carnation family)
Name Origin: From the Greek words *Di* (Zeus) and *anthos* (flower).
Species: *barbatus, caryophyllus, chinensis;* hybrids and cultivars
Common Names: sweet William (*D. barbatus*); carnation, clove pink *(D. caryophyllus);* pixie, spray carnation, mini-carnation *(D. caryophyllus* nana); pinks, Chinese pink, annual pink *(D. chinensis)*
Availability: Year-round
Description: Carnations are one of the floral industry's staples; available in a wide range of colors and varieties; white and other light colors can be color-dyed or tipped with paint; most are fragrant; versatile and offer many design uses. Sweet Williams are densely packed clusters of florets; available in pinks, reds, purples, and white; useful as a filler or accent. Miniature carnations have several buds and flowers on each stem; large variety of colors; work well as fillers or in adding mass. Carnations are used in corsages, boutonnieres, leis, and wedding designs.
Vase Life: Carnations and pixies are long lasting, 1 to 2 weeks; sweet William lasts 5 to 10 days; ethylene sensitive.

DIDISCUS see ***Trachymene***

DIGITALIS (see Plate 9)
(di-ji-TAL-lis)
Family: Scrophulariaceae (foxglove family)
Name Origin: From the Latin word *digitus* (finger), referring to the fingerlike florets.
Species: *purpurea, grandiflora (D. ambigua), lanata*
Common Name: foxglove
Availability: June through September
Description: Tall spikelike, often one-sided, racemes. The downward-facing tubular florets are splashed in the throat with exotic spots. Available in white and various shades of red, purple, pink, and yellow. Useful for adding line to arrangements.
Vase Life: Long lasting, about 2 weeks, with florets continously opening

DILL see ***Anethum***

DIOSMA
(die-OZ-mah)
Family: Rutaceae (citrus family)
Species: *ericoides;* sometimes referred to as *Coleonema ericoides*

GLORIOSA
gloriosa lily
GLADIOLUS
glad
GLADIOLUS
mini glad
ECHINOPS
globe thistle

DIOSMA
breath of heaven

Common Names: diosma, coleonema, breath of heaven
Availability: February through May
Description: Fragrant, heathlike shrubs with tiny, white, pink, lavender, or red flowers in terminal cymose clusters.
Vase Life: 5 to 7 days

DRYANDRA

(dry-AN-dra)
Family: Proteaceae (protea family)
Species: *floribunda, formosa;* others
Common Name: dryandra
Availability: Year-round
Description: Dense-headed flowers mostly available in yellow, but also available in oranges and reds, with prickly toothed leaves, similar to their *Banksia* relatives. These flowers provide accent and emphasis.
Vase Life: Long lasting, 2 to 3 weeks

E

EASTER LILY see ***Lilium***

ECHINOPS (see Plate 12)

(EK-i-nops)
Family: Compositae or Asteraceae (sunflower family)
Name Origin: From the Greek words *echinos* (hedgehog) and *-opsis* (appearance).
Species: *ritro, humilis*
Common Name: globe thistle
Availability: July through October
Description: Intense blue, white, or metalic blue, round, thistlelike flowers at the end of branching stems. The bold texture of these flowers creates striking effects.
Vase Life: Long lasting, 1 to 3 weeks

EREMURUS (see Plate 9)

(air-re-MOUR-us *or* e-ray-MEW-rus)
Family: Liliaceae (lily family)
Name Origin: From the Greek words *eremia* (desert) and *oura* (tail), referring to their desert habitat and the shape of the inflorescence.
Species: *robustus, stenophyllus*
Common Names: foxtail lily, desert candle, king's spear
Availability: May through September
Description: Tall spikelike racemes with small star-shaped florets on leafless stems. Generally yellow, but white, cream, orange, and pink varieties are also available. Excellent for adding line.
Vase Life: Long lasting, 1 to 3 weeks, during which time tiny new florets continuously open.

ERICA (see Plate 13)

(AIR-i-kah)
Family: Ericaceae (heath family)
Species: *canaliculata, codonodes* (Spanish heather), *persoluta, melanthera* (Scotch heather); others and hybrids. The genus *Calluna* (kah-LOO-nah) is known as common heather. Many other genera, such as *Thryptomene,* are heathlike shrubs and are often called heather or heath.
Common Names: heath, heather
Availability: November through April
Description: Spikelike flower clusters in panicles and racemes, with tiny bell-shaped florets. Available in pinks, purples, white, yellows, and green. Useful as a filler. Taller varieties can create line.
Vase Life: Long lasting, 1 to 3 weeks

CLARKIA
godetia

GYPSOPHILA
baby's breath

ERICA
heather

ERYNGIUM
sea holly

EUCHARIS
Amazon lily

ERYNGIUM

(air-IN-jee-um)
Family: Umbelliferae or Apiaceae (carrot family)
Species: *alpinum, amethystinum, giganteum;* others and hybrids
Common Names: sea holly, eryngo
Availability: June through October
Description: Flower head is cone-shaped, surrounded by feathery bracts. Available in silver-purple, blues, some pinks, green, and white. Adds unusual texture and form to arrangements. Well suited for many design styles.
Vase Life: Long lasting, 1 to 2 weeks

EUCHARIS

(YOU-kah-ris)
Family: Amaryllidaceae (daffodil family)
Species: *amazonica (E. grandiflora)*
Common Names: Amazon lily, eucharis lily, lily of the Amazon
Availability: Year-round
Description: three to six individual fragrant, large white flowers are borne in umbel patterns on leafless stems.
Vase Life: 7 to 14 days; individual flowers continue to open.

EUPHORBIA (see Plate 10)

(you-FOR-bee-ah)
Family: Euphorbiaceae (spurge family)
Name Origin: Named after Euphorbus, the physician to the king of Mauritania, Juba.
Species: *fulgens, pulcherrima, marginata;* others
Common Names: scarlet plume, spurge *(E. fulgans);* poinsettia, Christmas flower *(E. pulcherrima);* snow on the mountain, ghostweed *(E. marginata)*
Availability: September through February
Description: *E. fulgans* has small flowers alongside drooping stems. Adds colorful, graceful lines to designs. Available in orange, red, yellow, pink, and white. *E. marginata* with its green and white color pattern is useful as a foliage. All *Euphorbia* species bleed milky sap when cut or if leaves are removed; it can produce a severe dermatitis in susceptible individuals.
Vase Life: 7 to 10 days; condition in hot water with floral preservative.

EUSTOMA (see Plate 18)

(yew-STOW-mah)
Family: Gentianaceae (gentians)
Species: *grandiflorum (Lisianthus russellianus)*
Common Names: lisianthus, prairie gentian
Availability: May through December

HELICONIA
lobster claw

HELICONIA
torch heliconia

ANIGOZANTHOS
kangaroo paw

Description: Anemone-shaped flowers solitary on stems or in branching panicles. Generally three to five flowers open on each stem. Available in purples, pinks, reds, bicolors, and white. A variety of single, double, and triple petal forms are available; adds mass with a soft texture; single blooms work well for corsages.
Vase Life: Long lasting, 1 to 2 weeks

EVERLASTING FLOWER see ***Helichrysum***

F

FALSE DRAGONHEAD see ***Physostegia***
FALSE SPIRAEA see ***Astilbe***
FEVERFEW see ***Chrysanthemum***
FLAME TIP see ***Leucodendron***
FLAMINGO FLOWER see ***Anthurium***

FORSYTHIA (see Plate 10)

(for-SITH-ee-a)
Family: Oleaceae (olives, ashes, and lilacs)
Name Origin: Named after Scottish gardener, William Forsyth (1737–1804).
Species: *x intermedia, ovata*
Common Names: forsythia, golden bells
Availability: November through March
Description: Small yellow flowers clustered along woody stems, appearing before leaves. Tall stems add colorful line. Short stems work well as fillers. Useful in oriental and contemporary design styles as well as traditional springtime bouquets.
Vase Life: 1 to 2 weeks

FOXGLOVE see ***Digitalis***
FOXTAIL LILY see ***Eremurus***

FREESIA (see Plate 11)

(FREE-zee-ah *or* FREE-sha)
Family: Iridaceae (iris family)
Name Origin: Named after a German physician, Friedrich Freese (1800s).
Species: *x hybrida*
Common Name: freesia
Availability: Year-round
Description: Lovely, fragrant flowers; the florets are funnel shaped and branch asymmetrically off of a main stem on the upper side of a curved spike; available in a wide variety of colors; popular for wedding flowers and corsages. Interesting forms work well as an accent or filler in contemporary design styles.
Vase Life: 7 to 10 days; ethylene sensitive; hydrate well.

G

GAILLARDIA (see Plate 11)

(gay-LARD-ee-ah)
Family: Compositae or Asteraceae (sunflower family)
Name Origin: Named after French magistrate, Gaillard de Charentonneau (18th century).
Species: *aristata* cultivars
Common Name: blanket flower
Availability: July through September
Description: Solitary, large, showy, daisylike bicolored flowers in yellows, oranges, and reds. Bright coloration of these flowers creates emphasis. Gaillardia provide mass in mixed bouquets.
Vase Life: 5 to 7 days

GARDENIA (see Plate 11)

(gar-DEE-nee-ah)
Family: Rubiaceae (gardenia, coffee, and quinine)
Name Origin: Named after a Scottish physician and botanist, Alexander Garden (1730–1791), of South Carolina.
Species: *jasminoides* cultivars
Common Names: gardenia, cape jasmine
Availability: Year-round
Description: Fragrant, waxy, multipetaled flowers. Available in white to cream. Commercially they are packaged in special boxes wrapped for a high humidity; generally three to a box; each flower has a support collar made from thin cardboard and leaves. Gardenias bruise easily; moisten hands when working with them. Popular corsage and wedding flower.
Vase Life: Short lived, 1 to 2 days

GAYFEATHER see ***Liatris***
GENISTA see ***Cytisus***

GERBERA (see Plate 11)

(GER-be-rah)
Family: Compositae or Asteraceae (sunflower family)
Name Origin: Named after a German naturalist, Traugott Gerber (1700s).
Species: *jamesonii;* others
Common Names: gerbera, Transvaal daisy, African daisy, barberton daisy, veldt daisy
Availability: Year-round
Description: Daisylike flowers 3 to 5 inches across. Single and double-petaled forms. Available in a wide range of colors and bicolor patterns, usually with contrasting centers.

IXIA
corn lily
HYACINTHUS
hyacinth
MUSCARI
grape hyacinth
DAHLIA
IRIS

Stems are fleshy and leafless. These flowers have many uses. Can easily provide a focal point or add mass. Useful in traditional and contemporary design styles.
Vase Life: 3 to 7 days or longer; use clean water and containers to avoid plugged stems and bent neck; gerberas prefer preservative solution to the use of floral foam.

GERMAN STATICE see ***Limonium***
GIANT BOTTLEBRUSH see ***Banksia***
GILLYFLOWER see ***Matthiola***
GINGER see ***Alpinia***

GLADIOLUS (see Plate 12)
(gla-dee-OH-lus)
Family: Iridaceae (iris family)
Name Origin: From the Latin word for a small sword, referring to the shape of the leaves.
Species: hybrids
Common Names: gladiolus, glads, sword lily, corn flag
Availability: Year-round; smaller varieties available March through July
Description: Flowers are geotropic and are arranged on a thick stem in a one-sided spike or spikelike raceme. Available in a wide range of colors. Florets, texures, and shapes vary and may be ruffled, fringed, or plain and shaped like orchids, tulips, or roses. The florets of miniature gladiolus are more loosely arranged on the stem. Both are excellent line flowers. Individual florets are useful for corsage work.
Vase Life: 1 to 2 weeks; ethylene sensitive.

GLOBE AMARANTH see ***Gomphrena***
GLOBE ARTICHOKE see ***Cynara***
GLOBEFLOWER see ***Trollius***
GLOBE THISTLE see ***Echinops***

GLORIOSA (see Plate 12)
(glow-ree-OH-sah)
Family: Liliaceae (lily family)
Name Origin: From the Latin word *gloriosus* (glorious).
Species: *rothschildiana, superba*
Common Names: gloriosa lily, glory lily, climbing lily
Availability: Year-round
Description: Flowers have reflexed petals that are curled at the margins. Brightly colored red and yellow flowers on leafless stems. Unusual flower form provides emphasis. These flowers work well in oriental and contemporary design styles.
Vase Life: 5 to 10 days, with individual blooms continuously blooming, each lasting 4 or 5 days.

GLORIOSA DAISY see ***Rudbeckia***

GLORY LILY see ***Gloriosa***
GOAT'S BEARD see ***Astilbe***
GODETIA see ***Clarkia***
GOLDENROD see ***Solidago***
GOLDEN SHOWER see ***Oncidium***

GOMPHRENA
(gom-FREE-nah)
Family: Amaranthaceae (cockscombs and celosias)
Species: *globosa, haageana*
Common Name: globe amaranth
Availability: July through September
Description: Round flower heads about 1 inch across on the top of long stems. The flower head is made up of tiny fluffy flowers. Available in white, pinks, purples, and orange.
Vase Life: 5 to 10 days

GONIOLIMON see ***Limonium***
GRAPE HYACINTH see ***Muscari***
GUELDER ROSE see ***Viburnum***
GUERNSEY LILY see ***Nerine***

GYPSOPHILA (see Plate 13)
(jip-SOF-i-la *or* jip-so-PHIL-la)
Family: Caryophyllaceae (carnation family)
Name Origin: From the Greek words *gypsos* (gypsum) and *philos* (loving), referring to some species favoring gypsum or lime.
Species: *elagans, paniculata* (million stars); others and cultivars
Common Names: baby's breath, gyp
Availability: Year-round
Description: Fragrant complex panicles or dichasial cymes of white or pinkish white florets. A popular and delicate filler for arrangements. Small clusters work well in corsages, boutonnieres, and wedding flowers.
Vase Life: 5 to 7 days; ethylene sensitive; favors cool temperatures and high humidity.

H

HEATH/HEATHER see ***Erica***

HELIANTHUS (see Plate 26)
(hee-li-ANTH-us)
Family: Compositae or Asteraceae (sunflower family)
Name Origin: From the Greek words *helios* (sun) and *anthos* (flower).
Species: *annuus, tuberosus;* others and hybrids

CONSOLIDA larkspur

LIATRIS gay feather

SYRINGA lilac

LUPINUS lupine

Common Name: sunflower
Availability: Year-round, with peak supplies June thorugh October
Description: Large daisylike head flowers up to 10 inches across with petals (ray florets) surrounding a contrasting center (disc florets); single and double forms; available in creams, yellows, oranges, reds, and browns; provides mass and emphasis.
Vase Life: 7 to 10 days

HELICHRYSUM (see Plate 27)

(hee-li-KRIS-um)
Family: Compositae or Asteraceae (sunflower family)
Name Origin: From the Greek words *helios* (sun) and *chryson* (golden).
Species: *bracteatum;* others and cultivars
Common Names: strawflower, everlasting flower
Availability: July through September
Description: Brightly colored flowers with crisp, papery texture. Available in a wide range of colors including yellows, oranges, reds, pinks, and white. Sizes vary, most common about 2 inches across. Adds interesting texture and mass to arrangements.
Vase Life: Long lasting, 1 to 2 weeks

HELICONIA (see Plate 14)

(hel-i-KO-nee-ah)
Family: Musaceae or Heliconiaceae (heliconia family)
Species: *humilis, caribaea, psittacorum, pendula, chartacea, magnifica, xanthovillosa;* others
Common Names: heliconia, lobster claw, rainbow, parrot's flower; many common names with each species

HIPPEASTRUM
amaryllis

Availability: Year-round
Description: Tropical erect or drooping flowers; brilliant bracts of inflorescence densely packed together on thick stems. Most are bicolored in reds, oranges, yellows, and greens. Use alone or with other tropical flowers and foliage. Coloring and unique shape demand attention.
Vase Life: Long lasting, 10 to 14 days

HELIPTERUM

(hee-LIP-tur-um)
Family: Compositae or Asteraceae (sunflower family)
Name Origin: From the Greek words *helios* (sun) and *pteron* (wings), referring to their sun-loving nature and their feathery bristles.
Species: *humboldtianum, manglesii, roseum*
Common Names: Swan River everlasting, immortelles
Availability: May through October
Description: Daisylike, straw-textured head flowers on thin stems. Available in white and pink with yellow centers. Adds mass and also used as a filler.
Vase Life: 5 to 7 days

HELLEBORUS

(he-li-BOHR-us)
Family: Ranunculaceae (buttercup family)
Species: *niger, orientalis;* others and hybrids
Common Names: Christmas rose, Lenten rose, hellebore
Availability: Year-round
Description: Large and attractive white, greenish, pink, and purple flowers 2 to 5 inches across. Five petals surround golden anthers. Useful for adding emphasis and mass.
Vase Life: 5 to 7 days

HIPPEASTRUM

(hip-ee-AS-trum)
Family: Amaryllidaceae (daffodil family)
Name Origin: From the Greek word *hippos* (horse) referring to the inflorescence of the species *H. puniceum,* which was likened to the head of a horse.
Species: hybrids
Common Names: amaryllis, Barbados lily
Availability: Year-round
Description: Hippeastrum has four to six trumpet-shaped flowers clustered together in an umbel pattern at the top of a thick, hollow, leafless stem. Individual flowers are 5 to 9 inches across. Available in white, pinks, reds, oranges, and bicolors. These flowers have distinctive shapes that demand attention and work well in contemporary design styles. Allow space for blossoms to open.
Vase Life: 1 to 2 weeks, while individual florets continue opening.

HYACINTH see *Hyacinthus*

LILIUM—Many hybrids are available, including large and showy oriental hybrids and smaller flowered Asian hybrids.

HYACINTHUS (see Plate 15)

(hi-ah-SIN-thus)
Family: Liliaceae (lily family)
Species: *orientalis* cultivars
Common Names: hyacinth, Dutch hyacinth
Availability: November through April
Description: Compact spikelike racemes with bell-shaped, waxy florets. Available in white, pinks, reds, blues, purples, and yellows. Extremely fragrant. These flowers are at home with other bulb flowers in traditional spring designs. Also distincitve in contemporary designs, particularly vegetative and parallel styles.
Vase Life: 3 to 7 days

HYDRANGEA

(hi-DRAN-jee-ah)
Family: Saxifragaceae (hydrangea family)
Name Origin: From the Greek words *hydro* (water) and *aggos* (jar), referring to the fruits that are shaped like cups.
Species: *macrophylla, paniculata;* others
Common Name: hydrangea
Availability: July to October
Description: Large rounded or pyramidal compound clusters of small starlike florets in blues, pinks, and white. These flowers provide mass and emphasis in large arrangements.
Vase Life: 5 to 10 days; seal the bleeding latex for longer vase life.

HYDRANGEA

HYPERICUM

(hi-PEAR-i-kum *or* hi-pe-REE-kum)
Family: Guttiferae (mangosteen and mammey apple)
Name Origin: From the Greek words *hyper* (above) and *eikon* (a picture), it was hung above pictures to ward off evil spirits.
Species: *androsaemum* (Greek name for a plant with red sap, from *andros* [man] and *haima* [blood]); *hookerianum;* others and varieties
Common Names: hypericum berry, coffee bean berry, St. John's wort
Availability: June through December
Description: Most species produce bright yellow flowers on woody stems; their fruit or "berries" are their main attraction; berries grow in clusters and are brightly colored when ripe; berries are available in red, green, brown, and black; some varieties have reddish foliage.
Vase Life: 2 weeks; however if leaves are removed, stems will last 3 to 4 weeks.

***HYPERICUM* coffee bean berry**

I

IBERIS (see Plate 5)

(i-BEER-is)
Family: Cruciferae or Brassicaceae (mustard family)
Name Origin: From the Greek word *iberis* (from Iberia).
Species: *amara, umbellata;* others and cultivars
Common Names: candytuft, rocket candytuft
Availability: March through August
Description: Clustered florets that form a convex corymb or elongated raceme inflorescence. Available in white, pink, red, and lavender. Useful as a filler in mixed arrangements. Also works well grouped in contemporary design styles.
Vase Life: 5 to 7 days

INCA LILY see *Alstroemeria*

IRIS (see Plate 15)

(EYE-ris)
Family: Iridaceae (iris family)
Name Origin: Named after the Greek goddess of the rainbow.
Species: *reticulata, sibirica;* others and hybrids
Common Names: iris, Dutch iris, flag

NERINE
nerine lily
EUSTOMA
lisianthus
CONVALLARIA
lily of the valley
PHYSOSTEGIA
false dragonhead

Availability: Year-round, with peak supplies March through May
Description: Distinctive flower forms available in blues, purples, yellows, and white. Useful in a variety of design styles. Can be used for emphasis and accent. Allow plenty of space around flowers, as they continue to open. Striking in oriental and contemporary designs.
Vase Life: 2 to 6 days; ethylene sensitive; avoid water loss. Avoid warm temperatures, drafts, and sunlight to help flowers open.

IXIA (see Plate 15)

(IKS-ee-ah)
Family: Iridaceae (iris family)
Name Origin: From the Greek word *ixia* (bird lime), referring to the sticky sap.
Species: *viridiflora* hybrids
Common Names: corn lily, African corn lily
Availability: March through August
Description: Star-shaped florets in spike or panicle clusters on wiry stems. Available in cream, yellow, pink, orange, and red with mixtures of these colors. Works well as a line flower. Can also be used as a filler in larger bouquets. Especially striking in contemporary design styles.
Vase Life: 5 to 10 days, with individual florets continuously opening.

J

JONQUIL see *Narcissus*
JOSEPH'S COAT see *Amaranthus*

K

KANGAROO PAW see *Anigozanthos*
KISS OF DEATH see *Costus*

KNIPHOFIA

(nee-FOF-ee-ah)
Family: Liliaceae (lily family)
Name Origin: Named after Johann H. Kniphof (1704–1763).
Species: *uvaria* hybrids
Common Names: red hot poker, tritoma, torch lily, poker plant
Availability: June through October
Description: Tight, terminal, spikelike racemes packed with overlapping florets of red, orange, and yellow on leafless stems. The scarlet florets become orange and then yellow with age. These tall-stemmed flowers add emphasis and height and look good in contemporary design styles.
Vase Life: 7 to 10 days, with florets continuously opening.

KNIPHOFIA
red hot poker

L

LADY'S MANTLE see *Alchemilla*
LADY'S SLIPPER see *Paphiopedilum*
LARKSPUR see *Consolida*

LATHYRUS (see Plate 26)

(LATH-i-rus)
Family: Leguminosae or Fabaceae (pea family)
Name Origin: The Greek name for the pea.
Species: *odoratus* cultivars
Common Name: sweet pea
Availability: February through September
Description: Sweetly scented flowers with delicate petals in soft colors, including white, pinks, reds, blues, and lavenders. Stems are fairly short with three to seven flowers per stem. Sweet peas can be used to add mass or used as a filler. Simple and charming alone or with other spring and summer flowers.
Vase Life: 3 to 7 days; ethylene sensitive.

LAVATERA

(la-vah-TER-ah)
Family: Malvaceae (cotton, mallows, and hollyhocks)
Name Origin: Named after Zurich naturalists, the Lavater brothers of the 16th century.
Species: *trimestris*
Common Names: mallow, tree mallow
Availability: June to October
Description: Trumpet-shaped flowers that look like miniature Hawaiian hibiscus flowers. Several flowers appear at

Members of the* Orchidacea *family—Many genera, species, and hybrids.

the top of tall leafy stems. Available in reds, pinks, lavendar, and white.
Vase Life: 5 to 10 days

LEI ORCHID see *Vanda*

LEPTOSPERMUM (see Plate 22)
(lep-toe-SPUR-mum)
Family: Myrtaceae (myrtles, eucalyptus, and cloves)
Name Origin: From the Greek words *leptos* (slender) and *sperma* (seed), referring to its narrow seeds.
Species: *scoparium*
Common Names: lepto, tea tree, New Zealand tea tree
Availability: Peak supplies January through April
Description: Clusters of small blossoms on woody stems. Available in white, pinks, and reds. Long branches are excellent for adding line. Smaller stems work well as spiky fillers. Provides an elegant line in oriental and contemporary design styles.
Vase Life: Long lasting, 7 to 10 days

LEUCODENDRON (see Plate 20)
(loo-ka-DEN-dron)
Family: Proteacea (protea family)
Name Origin: From the Greek words *leukos* (white) and *dendron* (tree), referring to its silvery foliage.
Species: *argenteum;* others and cultivars
Common Names: silver tree, flame tip
Availability: Year-round; some species are limited by season.
Description: Flower head consists of stiff bracts surrounding a cone or small inconspicuous flower. Male flowers have terminal, sessile heads; female flowers have terminal, cone-like heads. The woody bracts, often mistaken for petals, are generally colorful, most often reds, burgundy, green, and yellows, and combinations of these colors. The stems are densely packed with stiff leaves that radiate out on all sides. These flowers work well with other exotic flowers.
Vase Life: Long lasting, 2 to 3 weeks

LEUCOSPERMUM (see Plate 20)
(lu-co-SPER-mum)
Family: Proteaceae (protea family)
Species: *cordifolium, reflexum, nutans, catherinae;* others and cultivars
Common Name: pincushion protea
Availability: Year-round; some species are limited by season.
Description: The flower heads consist of colorful orange or reddish orange styles that form a domed, globular shape (thus the name pincushion). Flowers top woody stems loaded with stiff leaves. Creates emphasis through texture and form.
Vase Life: Long lasting, 2 to 3 weeks

LIATRIS (see Plate 16)
(lie-AH-tris)
Family: Compositae or Asteraceae (sunflower family)
Species: *spicata; callilepsis,* and others
Common Names: liatris, gay feather, blazing star, button snakeroot, purple poker
Availability: Year-round
Description: Tiny florets appear in dense spikes; opening from the top downward. Available in purple and white. Excellent for creating strong line in arrangements. Works well in mixed bouquets and is striking in oriental and contemporary design styles.
Vase Life: 7 to 10 days

LILAC see *Syringa*

LILIUM (see Plate 17)
(LIL-ee-um)
Family: Liliaceae (lily family)
Name Origin: The Latin name for these bulbous flowers.
Species: *longiflorum;* others, hybrids, and cultivars
Common Names: Lilies bred from *L. auratum* and *L. speciosum* (such as *L.S.* 'Rubrum') are large and called oriental lilies and are generally white, pink, and white with red. Asian lilies are generally the yellows, oranges, reds, and whites; Easter lily (*L. longiflorum)* is white and trumpet-shaped.
Availability: Year-round
Description: The flowers are 4 to 6 inches across. A wide range of forms and colors are available. Flowers appear on short branchs at the end of tall stems. Some are upright, while others are pendant or outward-facing. The striking form of lilies with their central floral parts provides a focal point. Leave room for lily blossoms to open. Lilies work well in mixed arrangements, wedding, and sympathy designs. Striking in oriental and contemporary designs.
Vase Life: 4 to 5 days per bloom; ethylene sensitive; remove anthers to prevent staining.

LILY see *Lilium*
LILY OF THE FIELD see *Anemone*
LILY OF THE INCAS see *Alstroemeria*
LILY OF THE NILE see *Agapanthus*
LILY OF THE VALLEY see *Convallaria*

LIMONIUM (see Plate 25)
(lee-MO-nee-um)
Family: Plumbaginaceae (see lavender and thrift)
Name Origin: From the Greek word *leimon* (meadow), referring to their natural habitat.
Species: *latifolium, perezii, sinuatum, ferulaceum* (caspia and misty blue hybrids); others and cultivars

Members of the* Proteaceae *family.

Common Names: statice, sea lavender, caspia, misty blue, seafoam statice; *Goniolimon (L. tataricum)* (German statice)
Availability: Year-round
Description: Tiny white or yellow flowers are surrounded by papery bracts. Flowers are sessile, in panicles or spikes. Bract colors include white and various shades and tints of pink, yellow, blue, and purple. Useful as a colorful filler adding interesting texture to mixed bouquets. Misty blue and caspia are sometimes odoriferous.
Vase Life: Long lasting, 1 to 2 weeks; subject to mildew and brown spots; provide air circulation.

***LIMONIUM* misty blue**

LISIANTHUS see ***Eustoma***
LOBSTER CLAW see ***Heliconia***
LOOSESTRIFE see ***Lysimachia***
LOVE IN A MIST see ***Nigella***
LOVE LIES BLEEDING see ***Amaranthus***
LUPINE see ***Lupinus***

LUPINUS (see Plate 16)
(lu-PEEN-us)
Family: Leguminosae or Fabaceae (pea family)
Species: *polyphyllus* cultivars
Common Name: lupine
Availability: July through September
Description: Lupine florets resemble pea blossoms and are densely packed on tall, erect spikelike raceme stalks. Available in a wide variety of colors. These are graceful, old-fashioned flowers that look lovely in mixed summer bouquets.
Vase Life: 5 to 10 days

LYSIMACHIA
(lie-si-MAK-ee-ah)
Family: Primulaceae (primrose family)
Name Origin: Named after King Lysimachos of ancient Thrace. According to legend, the king was able to pacify a bull by using a piece of loosestrife.
Species: *clethroides, punctata*
Common Name: loosestrife
Availability: July to September
Description: Tiny white or yellow, star-shaped florets tightly arranged in small slender, curving, spikelike racemes at the top of dense foliage.
Vase Life: 5 to 7 days

M

MADAGASCAR JASMINE see ***Stephanotis***

MAGNOLIA
(mag-NOL-ee-ah)
Family: Magnoliaceae (Magnolia family)
Name Origin: Named after French professor of botany Pierre Magnol (1638-1715).
Species: many species and hybrids
Common Name: magnolia
Availability: February to May
Description: Elegant focal flowers on woody stems. Available in a variety of forms and colors; beautiful in oriental designs.
Vase Life: 3 to 6 days

MALLOW see ***Lavatera***
MARGUERITE DAISY see ***Chrysanthemum***
MARIGOLD see ***Tagetes***
MASTERWORT see ***Astrantia***
MATRICARIA see ***Chrysanthemum***

MATTHIOLA (see Plate 25)
(ma-THEE-oh-lah)
Family: Cruciferae or Brassicaceae (mustard family)
Name Origin: Named after Italian botanist Pierandrea Mattioli (1500–1577).
Species: *incana* and hybrids
Common Names: stock, gillyflower
Availability: January through October
Description: Small, 1-inch florets form rounded, spikelike racemes. Available in a wide selection of colors. These fragrant flowers add mass and line to mixed bouquets.
Vase Life: 3 to 7 days

MAGNOLIA

AMMI
Queen Anne's lace
PHLOX
ANANAS
ornamental pineapple
RANUNCULUS
Persian buttercup

MEADOWSWEET see *Astilbe*
MICHAELMAS DAISY see *Aster*
MILKWEED see *Asclepias*
MIMOSA see *Acacia*
MISTY BLUE see *Limonium*

MOLUCCELLA (see Plate 4)
(mahl-you-SEL-ah)
Family: Labiatae or Lamiaceae (mint family)
Species: *laevis*
Common Names: bells of Ireland, shellflower, molucca balm
Availability: Year-round, with peak supplies June through October
Description: Whorls of tiny, white, fragrant flowers are surrounded by curious green shell-like calyces resembling bells, which are often mistaken for petals. The flowers and sepals are clustered along tall stems. Often reserved for St. Patrick's Day, but useful throughout the year for adding line and accent. Striking in all-foliage and contemporary designs.
Vase Life: 7 to 10 days

MONKSHOOD see *Aconitum*
MONTBRETIA see *Crocosmia*
MONTE CASSINO see *Aster*
MOTH ORCHID see *Phalaenopsis*

MUSCARI (see Plate 15)
(mus-KAIR-ree)
Family: Liliaceae (lily family)
Species: *armeniacum, botryoides;* others
Common Name: grape hyacinth
Availability: January through March
Description: Tiny, tubular white or blue florets densely clustered on short stems. These small and delicate flowers work best in contempoary design styles, such as parallel systems or vegetative, where they can be clustered together for greater impact.
Vase Life: 4 to7 days

N

NARCISSUS (see Plate 7)
(nar-SIS-us)
Family: Amaryllidaceae (daffodil family)
Name Origin: Narcissus was a legendary youth of Greek mythology who was arrogant and fell in love with his own reflection.
Species: *pseudonarcissus;* others and hybrids
Common Names: narcissus, daffodil, jonquil
Availability: November through April
Description: The single, trumpet-shaped flowers are known as daffodils; other varieties are known as jonquils or narcissus. The tiny white narcissus in clusters are known as paperwhites. Narcissus flowers are available in a great range of forms, sizes, and colors—white, cream, yellows, oranges, and bicolors. Their form adds accent and emphasis to simple and contemporary design styles.
Vase Life: 4 to 6 days; condition narcissus alone, as they secrete sap when cut that is harmful to other flowers; do not recut again when mixing with other flowers in designs.

NERINE (see Plate 18)
(near-REEN)
Family: Amaryllidaceae (daffodil family)
Name Origin: Named after a Greek sea nymph.
Species: *bowdenii, sarniensis;* others
Common Name: Guernsey lily
Availability: Year-round
Description: Small florets are lilylike and clustered at the top of leafless stems in umbel patterns. There are from six to twelve florets forming each cluster. Available in pinks, reds, and white. These flowers work well in many design styles including mixed bouquets, oriental, and contemporary. The individual florets are useful in corsages, boutonnieres, and wedding designs. Odoriferous.
Vase Life: Long lasting, 1 to 2 weeks

NIGELLA
(nee-JEL-lah)
Family: Ranunculaceae (buttercup family)
Species: *damascena;* others

NIGELLA
love in a mist

Flowering Branches

Common Names: love in a mist, fennel flower, wild fennel
Availability: June to September
Description: Nigella flowers are small and flat; several appear at the top of a thin, hairy stem. Egg-shaped seed capsules among the flowers provide accent. Flowers are available in blue, pink, and white. Nigella provides interesting texture.
Vase Life: 5 to 7 days

O

OBEDIENT PLANT see ***Physostegia***

ONCIDIUM (see Plate 19)

(ahn-SID-ee-um)
Family: Orchidaceae (orchid family)
Name Origin: From the Greek word *onkos* (tumor), referring to a swelling on the lip.
Species: *pulchellum, splendidum, varicosum;* others and hybrids
Common Names: oncidium, golden shower, butterfly orchid, dancing doll, dancing lady orchid
Availability: Year-round
Description: Masses of tiny orchid flowers in thin, branching stems. The flowers are yellow with speckles of orange, red, and brown. The stems arch with the weight of the florets; provides graceful linear curves in arrangements. These flowers work well with other orchids and exotic flowers. Arching stems are beautiful in oriental designs.
Vase Life: Long lasting, 1 to 2 weeks

ONION FLOWER see ***Allium***
ORCHID see ***Cattleya, Cymbidium, Dendrobium, Oncidium, Paphiopedilum, Phalaenopsis, Vanda***
ORIENTAL LILY see ***Lilium***
ORNAMENTAL PINEAPPLE see ***Ananas***

ORNITHOGALUM (see Plate 27)

(or-ni-THAHG-ah-lum)
Family: Liliaceae (lily family)
Name Origin: From the Greek words *ornis* (bird) and *gala* (milk).
Species: *arabicum, thyrsoides;* others
Common Names: star of Bethlehem, chincherinchee
Availability: Year-round
Description: White, star-shaped flowers in tight racemes and corymb clusters at the top of leafless stems. Flower centers are either whitish green or black. Star of Bethlehem is a versatile flower, adding line, mass, or accent. Also useful as a spiky filler. A beautiful, long-lasting addition to mixed bouquets, oriental designs, and contemporary styles.
Vase Life: Long lasting, 2 to 3 weeks

OSTRICH PLUME see ***Alpinia***

OUTDOOR GYPSOPHILA see ***Saponaria***
OXEYE DAISY see ***Chrysanthemum***

P

PAEONIA (see Plate 24)

(pee-OH-ne-ah)
Family: Paeoniaceae or Ranunculaceae (peony family)
Name Origin: From the Greek name *paionia,* referring to Paion, who was the physician to the gods.
Species: *lactiflora, suffruticosa*
Common Name: peony
Availability: May through July
Description: Large, fragrant, single flowers from 3 to 8 inches across. Available in a wide variety of forms, including single, double, and anemone types. These giant blossoms are grand all by themselves or in mixed arrangements.
Vase Life: 3 to 7 days

PAINTED TONGUE see ***Anthurium***
PAINTER'S PALETTE see ***Anthurium***

PAPAVER

(pah-PAY-ver)
Family: Papaveraceae (poppy family)
Species: *orientale, nudicaule;* others and hybrids
Common Names: poppy, Iceland poppy
Availability: May through September

PAPAVER
poppy

ROSA—Many hybrids and cultivars.

Description: Papery petals surround a contrasting center; open up almost flat. Flowers are solitary at the top of nodding, wiry stems. Available in a variety of intense colors.
Vase Life: 1 to 5 days; to increase vase life, sear the stem ends with a flame or dip in boiling water and place in warm preservative solution.

PAPERWHITES see *Narcissus*

PAPHIOPEDILUM (see Plate 19)

(paf-ee-oh-PED-il-lum)
Family: Orchidaceae (orchid family)
Name Origin: From the Greek words *Paphos* the site of a temple on Cyprus where Aprhodite was worshipped and *pedilon* (slipper).
Species: *bellatulum, fairrieanum;* others and hybrids; greenhouse Cypripediums belong to the genera Paphiopedilum.
Common Names: slipper orchid, lady's slipper
Availability: Year-round
Description: Exotic flowers with dramatic form and coloring. Ideal for contemporary designs.
Vase Life: 5 to 7 days

PEONY see *Paeonia*
PEPPERBERRY see *Schinus*
PERSIAN BUTTERCUP see *Ranunculus*
PERUVIAN LILY see *Alstroemeria*

PHALAENOPSIS (see Plate 19)

(fal-en-NOP-sis)
Family: Orchidaceae (orchid family)
Name Origin: From the Greek words *phalaina* (moth) and *-opsis* (resembling).
Species: *amabilis, gigantea;* others and hybrids
Common Name: moth orchid
Availability: Year-round
Description: These delicate, elegant orchid flowers are 3 to 4 inches across and flatter-looking than most orchids. Available mostly in white and pink. Generally sold individually. Frequently used in wedding and corsage designs.
Vase Life: 3 to 5 days

PHLOX (see Plate 21)

(floks)
Family: Polemoniaceae (phlox family)
Name Origin: From the Greek word *phlox* (flame).
Species: *paniculata, drummondii*; others
Common Name: phlox
Availability: June through November
Description: Small florets form dense terminal panicles on leafy stems. Available in white, pinks, reds, lavenders, and bicolors. An excellent filler flower.
Vase Life: 3 to 7 days

PHYSOSTEGIA (see Plate 18)

(fie-soe-STEE-jee-ah)
Family: Labiatae or Lamiaceae (mint family)
Species: *virginiana* cultivars
Common Names: false dragonhead, obedient plant, lion's heart, obedience
Availability: July to October
Description: Tubular florets form along a 12- to 24-inch stem, forming a spikelike inflorescence with an unusual shape. Available in white, pinks, and lavender.
Vase Life: 1 week

PINCUSHION FLOWER see *Scabiosa*
PINCUSHION PROTEA see *Leucospermum*
PINEAPPLE, ORNAMENTAL see *Ananas*
PINKS see *Dianthus*
PITCHER PLANT see *Sarracenia*
PIXIE see *Dianthus*
PLUME THISTLE see *Cirsium*
POINSETTIA see *Euphorbia*

POLIANTHES (see Plate 28)

(pah-lee-ANTH-eez)
Family: Agavaceae (Mexican lily family)
Name Origin: From the Greek words *polis* (grey) and *anthos* (flower).
Species: *tuberosa*
Common Name: tuberose
Availability: February to October
Description: Clustered spikes of fragrant, waxy, white flowers. Star-shaped florets are about 1 inch across. These flowers can add mass, line, and accent to a number of design styles. Useful in corsages, leis, and wedding designs.
Vase Life: Long lasting, 1 to 2 weeks

POPPY see *Papaver*
POT MARIGOLD see *Calendula*
PRINCE'S FEATHER see *Amaranthus*

PROTEA (see Plate 20)

(PRO-tee-a)
Family: Proteaceae (protea family)
Name Origin: Named after the Greek sea God, *Proteus*, who had the power of prophecy.
Species: *compacta, cynaroides, grandiceps, magnifica, nerifolia, obtusifolia, repens;* others and hybrids
Common Names: protea; king protea *(P. cynaroides);* queen protea *(P. magnifica);* pink mink *(P. nerifolia);* many other common names
Availability: Year-round

SCABIOSA
pincushion flower
CARTHAMUS
safflower
PAEONIA
peony
ANTIRRHINUM
snapdragon

Description: Proteas are large focal flowers. They are rounded with colorful bracts in a variety of colors, forms, and textures. They stand singly at the top of woody, leafy stems.
Vase Life: Long lasting, 2 to 3 weeks

PRUNUS (see Plate 22)

(PRU-nus)
Family: Rosaceae (rose family)
Name Origin: *Prunus* is the Latin name for the plum tree.
Common Names: Prunus includes cherry, plum, peach, apricot, nectarine, and almond trees and shrubs.
Availability: October through May, depending on species. Can be forced to blossom indoors.
Description: Single or double flowers along woody branches. A wide variety of forms, depending on the species. Available in white and tints and shades of pink. Many types are fragrant. Tall branches add line. Shorter stems add linear fillers. Blossoming branches are elegant in oriental and vegetative designs.
Vase Life: Long lasting, 1 to 2 weeks

PUSSY WILLOW see *Salix*

Q

QUEEN ANNE'S LACE see *Ammi*
QUINCE see *Chaenomeles*

R

RAINBOW see *Heliconia*

RANUNCULUS (see Plate 21)

(rah-NUN-kew-lus)
Family: Ranunculaceae (buttercup family)
Name Origin: From the Latin word *rana* (frog), as many prefer to grow in wet areas.
Species: *asiaticus;* others and hybrids
Common Names: ranunculus, Persian buttercup
Availability: January through May
Description: These single and double flowers resemble small peonies, with centers often a contrasting color or black. Sizes vary from 1 to 4 inches wide. Available in a wide selection of bright colors including white, yellow, orange, red, and pink. Colorful additions to mixed bouquets and contemporary design styles.
Vase Life: 3 to 7 days

RED HOT POKER see *Kniphofia*

ROSA (see Plate 23)

(RO-za)
Family: Rosaceae (rose family)
Species: hybrids and cultivars
Common Name: rose
Availability: Year-round
Description: Roses are available in a wide range of colors and sizes. The flowers are solitary, corymbose, or panicled on erect stems. The heads are densely crowded with petals. Roses can be grouped simply by head size into standard or tea, large, medium, mini or sweethearts, and spray roses. The stem lengths are classified as shorts, medium, long, and extra long. Roses are popular cut flowers and are useful in many design styles. Excellent flowers for corsages and boutonnieres.
Vase Life: Varies greatly, from 3 to 14 days; somewhat ethylene sensitive; use a hydration solution.

ROSE see *Rosa*

RUDBECKIA (see Plate 4)

(rude-BEK-ee-ah)
Family: Compositae or Asteraceae (sunflower family)
Name Origin: Named after Olof Rudbeck the elder (1630–1702) and the younger (1660–1740).
Species: *fulgida, hirta;* others and hybrids
Common Names: coneflower, black-eyed Susan, gloriosa daisy
Availability: July through September
Description: Daisylike gold and orange flowers with prominant black, cone-shaped centers. These flowers add mass and accent. They are striking additions to mixed summer bouquets.
Vase Life: 7 to 10 days

S

SAFFLOWER see *Carthamus*

SALIX (see Plate 22)

(SAY-liks)
Family: Salicaceae (aspens, poplars, and willows)
Species: *discolor;* others
Common Names: *discolor* (pussy willow); *S. matsudana* 'Tortuosa' (corkscrew or curly willow); others
Availability: Fresh pussy willow is available January through April; curly willow is available year-round.

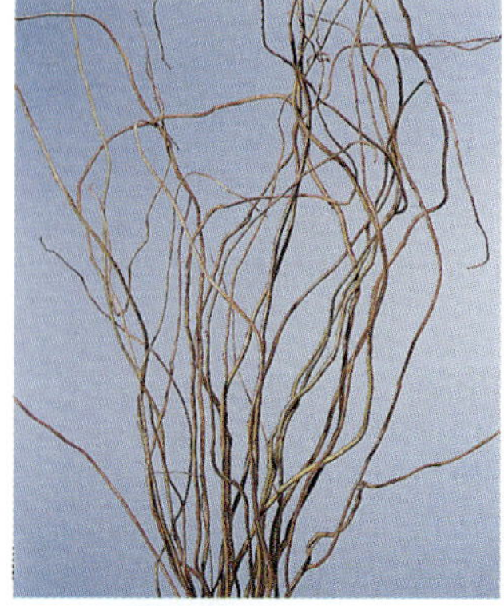

SALIX curly willow

LIMONIUM
sea lavender
statice
STEPHANOTIS
LIMONIUM
statice
MATTHIOLA
stock
LIMONIUM
German statice

Description: Pussy willow catkins are soft and fuzzy; they are greyish white and closely spaced along woody stems. Curly willow adds exciting lines to floral designs.
Vase Life: Long lasting, 10 to 14 days and longer; dries well

SALTBUSH see *Atriplex*

SAPONARIA

(sap-oh-NAH-ree-ah)
Family: Caryophyllaceae (carnation family)
Name Origin: From the Latin word *sapo* (soap). A soap can be made from the species *S. officinalis*. Saponaria is also known as *Vaccaria.*
Species: *officinalis;* others
Common Names: saponaria, bouncing bet, soapwort, outdoor gypsophila
Availability: June through September
Description: Small star-shaped flowers at the tips of cymose or paniculate branching stems. Available in white and pink. A delicate and airy filler in mixed arrangements.
Vase Life: 7 to 10 days

SARRACENIA (see Plate 26)

(sa-ra-SEE-nee-ah)
Family: Sarraceniaceae (pitcher plants)
Name Origin: Named after a French botanist and physician, Michael Sarrasin (1659–1734).
Species: *flava, purpurea;* others and hybrids
Common Names: sarracenia, pitcher plant, swamp lily, cobra head, cobra lily, trumpet
Availability: April through September
Description: Unusual curiosities that have unique form and venation color patterns. Often mistaken for flowers, these leaves with an odd tubular, trumpet shape provide emphasis, line, and accent.
Vase Life: 7 to 10 days

SAPONARIA
outdoor gypsophila

SATIN FLOWER see *Clarkia*

SAXICOLA see *Thryptomene*

SCABIOSA (see Plate 24)

(skab-ee-OH-sah)
Family: Dipsacaceae (teasel and scabious)
Name Origin: From the Latin word *scabies* (itch), for the rough leaves were said to cure the itch.
Species: *altropurpurea, caucasica*
Common Names: pincushion flower, scabiosus
Availability: June through October
Description: Round, single flowers with papery texture. Flowers are 2 to 3 inches across. Available in a wide variety of colors. Useful as a mass flower in mixed bouquets. The unusual texture and intricate circular form also can create a focal point in contemporary design styles.
Vase Life: 5 to 7 days

SCARLET PLUME see *Euphorbia*

SCHINUS

(SKY-nus *or* SHY-ness)
Family: Anacardiaceae (cashew, mango, sumacs, and poison ivy)
Species: *molle;* others
Common Names: pepperberry, California pepperberry, pepper tree
Availability: September through February
Description: Small red berries fill cascading and upright woody branches; adds vibrant color; good for Christmas designs; useful in a wide variety of designs.
Vase Life: Long lasting, 3 to 4 weeks; recut stems under water; do not smash stems; use hydrating solution; refrigerate until needed; dries well if hung upside down.

SCHINUS
pepperberry

SARRACENIA
swamp lily

LATHYRUS
sweet pea

HELIANTHUS
sunflower

SCILLA
bluebells

SCILLA
(SKIL-lah *or* SIL-lah)
Family: Liliaceae (lily family)
Name Origin: From the Greek name for the sea squill.
Species: *sibirica;* others and hybrids
Common Names: squill, bluebells, wood hyacinth
Availability: March through June
Description: Several small blue florets in racemes cluster the top of short stems. Works well in vegetative designs.
Vase Life: 7 to 10 days

SEA HOLLY see *Eryngium*
SEA LAVENDER see *Limonium*

SEDUM (see Plate 27)
(SEE-dum)
Family: Crassulaceae (stonecrops and houseleeks)
Name Origin: From the Latin word *sedo* (sit), a classical name for many succulent plants.
Species: *spectabile, telphium;* others and hybrids
Common Names: sedum, stonecrop
Availability: April through October
Description: Tiny star-shaped flowers that form dense terminal panicles. Available in yellow, pinks, reds, and white. Sedum works well as a filler in mixed bouquets.
Vase Life: 7 to 10 days

SHAMPOO GINGER see *Zingiber*
SHASTA DAISY see *Chrysanthemum*
SHELLFLOWER see *Moluccella*
SILVER TREE see *Leucodendron*
SINGAPORE ORCHID see *Dendrobium*
SLIPPER ORCHID see *Paphiopedilum*
SNAPDRAGON see *Antirrhinum*
SNOWBALL see *Viburnum*
SNOWBERRY see *Symphoricarpos*
SNOW ON THE MOUNTAIN see *Euphorbia*

SOLIDAGO (see Plate 27)
(so-li-DAY-go)
Family: Compositae or Asteraceae (sunflower family)
Name Origin: From the Latin word *solido* (to strengthen or make whole), making reference to its medicinal properties.
Species: *canadensis;* others
Common Names: solidago, goldenrod
Availability: May through October
Description: Tiny yellow flowers forming soft panicled or racemed plumes. Solidago works well as a filler.
Vase Life: 7 to 10 days

x SOLIDASTER (see Plate 27)
(so-li-DAS-ter)
Family: Compositae or Asteraceae (sunflower family)
Species: Solidaster is an intergeneric hybrid, from the names of the parents: *Aster* and *Solidago; S. luteus;* hybrids
Common Name: solidaster
Availability: Year-round
Description: Tiny yellow flowers on branching stems. Adds a fluffy texture. Excellent filler flower.
Vase Life: 7 to 10 days

SPIRAL GINGER see *Costus*
SPURGE see *Euphorbia*
SQUILL see *Scilla*
STAR OF BETHLEHEM see *Ornithogalum*
STATICE see *Limoniuim*

STEPHANOTIS (see Plate 25)
(ste-fa-NO-tis)
Family: Asclepiadaceae (milkweeds and waxplant)
Name Origin: From the Greek words *stephanos* (crown) and *otos* (ear). Greek name for myrtle, which was used to make crowns.
Species: *floribunda*

x SOLIDASTER
SOLIDAGO
goldenrod
HELICHRYSUM
strawflower
SEDUM
stonecrop
ORNITHOGALUM
star of Bethlehem

Common Names: stephanotis, steph, Madagascar jasmine
Availability: Year-round
Description: Small white fragrant flowers; tubular, star-shaped, waxy blossoms; flowers are cut off vines and sold stemless. Packaged in air-tight, humid boxes or bags. Stephanotis works well in corsages, boutonnieres, and wedding designs.
Vase Life: 1 to 4 days

STOCK see *Matthiola*
STONECROP see *Sedum*
STRAWFLOWER see *Helichrysum*

STRELITZIA (see Plate 2)

(stre-LITS-ee-a)
Family: Strelitziaceae or Cannaceae (bird of paradise flower)
Name Origin: Named after Queen Charlotte of Mecklenberg-Strelitz (1744–1818).
Species: *reginae*
Common Names: bird of paradise, crane lily; giant bird of paradise *(S. nicolai)*
Availability: Year-round
Description: Showy blossoms are oddly shaped, resembling a bird; rigid, boatlike bracts; stems are tall and thick. These tropical flowers demand attention; use alone or with other exotic flowers and foliage.
Vase Life: 1 to 2 weeks; beauty and longevity of flowers can be increased by gently lifting out florets.

SUMMER TULIP see *Curcuma*
SUNFLOWER see *Helianthus*
SWAMP LILY see *Sarracenia*
SWEET BROOM see *Cytisus*
SWEET PEA see *Lathyrus*

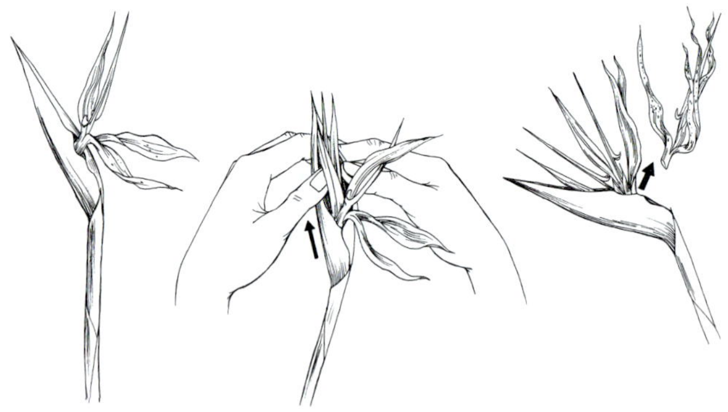

***STRELITZIA* bird of paradise**

The individual bird of paradise florets can gently be lifted out of the bract, by using wet hands, increasing longevity and enhancing the flower's unique beauty. As shown at right, old and bruised florets can be removed.

SWEET WILLIAM see *Dianthus*
SWORD LILY see *Gladiolus*

SYMPHORICARPOS

(sim-fo-ree-KAR-pos)
Family: Caprifoliaceae (elders and honeysuckles)
Name Origin: From the Greek words *symphroein* (bear together) and *karpos* (a fruit) referring to the clustered fruits.
Species: *albus, orbiculatus*
Common Names: snowberry; *(S. orbiculatus* is called Indian currant and coral berry)
Availability: August through November
Description: Fine and twiggy branches with tight berry clusters; white and pink berries.
Vase Life: 2 or more weeks; remove leaves to highlight the berries; leaves are susceptible to powdery mildew; cut stems under water; use preservative solution; change water as needed.

SYRINGA (see Plate 16)

(si-RIN-gah)
Family: Oleaceae (lilac family)
Name Origin: From the Greek word *syrinx* (pipe), referring to the hollow stems.
Species: *vulgaris;* others and cultivars
Common Name: lilac
Availability: December through May
Description: Clustered florets in a compound dichasium inflorescence. Individual florets are star-shaped and available in white, cream, pinks, and purples. Most varieties are fragrant.
Vase Life: Varies greatly, from 2 to 10 days; remove all foliage from woody stems for increased cut life. Lilacs also perform better when arranged in preservative solution without the use of floral foam. Recut stem ends often.

T

TAGETES

(tah-JEE-teez)
Family: Compositae or Asteraceae (sunflower family)
Name Origin: From Tages, an Etruscan deity, the grandson of Jupiter, who sprang from the newly plowed earth.
Species: *erecta, patula;* others and hybrids
Common Names: marigold; African marigold *(T. erecta);* French marigold *(T. patula)*
Availability: July through September
Description: Bright orange and yellow daisylike or carnation-like flowers. Useful for adding mass to mixed summer bouquets. Most have a unique fragrance.
Vase Life: 1 to 2 weeks

TAIL FLOWER see *Anthurium*

TULIPA
tulip

POLIANTHES
tuberose

CHAMELAUCIUM
waxflower

TAGETES
marigold

TELOPEA

(tay-LO-pee-ah)
Family: Proteaceae (protea family)
Name Origin: From the greek word *telopos* (seen from afar), referring to the showy flowers.
Species: *speciosissima*
Common Name: Waratah
Availability: May through November
Description: Dense terminal racemes surrounded by colored bracts. The flowers are vibrant red and add dramatic emphasis to designs.
Vase Life: 1 to 2 weeks

THOROUGHWAX see ***Bupleurum***

THROATWORT see ***Trachelium***

THRYPTOMENE

(thrip-toe-MEAN-ee *or* THRIP-toe-mine)
Family: Myrtaceae (myrtles, eucalyptus, and cloves)
Species: *saxicola, calycina;* others
Common Names: saxicola, calycina and calcinia, Australian heather, mini-waxflower
Availability: November through April
Description: Heatherlike spikes of tiny pink or white flowers on woody stems; good linear filler; resembles waxflower.
Vase Life: 3 to 7 days

THRYPTOMENE saxicola

TORCH LILY see ***Kniphofia***

TRACHELIUM

(tra-KEEL-lee-um)
Family: Campanulaceae (bellflower family)
Name Origin: From the Greek word *trachelos* (neck), referring to supposed medicinal properties.
Species: *caeruleum*
Common Names: throatwort, blue throatwort
Availability: March through November
Description: Clustered flowers in dense terminal compound corymbs; available in pink, purple, blue, and white. Excellent for adding mass or can be used as a filler in large bouquets. Entire flower heads can be used for basing contemporary design styles.
Vase Life: 7 to 10 days

TELOPEA
waratah

TRACHELIUM throatwort

TRACHYMENE didiscus

TRACHYMENE

(tray-ki-MEE-nee)
Family: Umbelliferae or Apiaceae (carrot family)
Name Origin: From the Greek words *trachys* (rough) and *meninx* (membrane), referring to the fruit.
Species: *caerulea (Didiscus caeruleus)*
Common Names: didiscus, blue laceflower, laceflower
Availability: June through November
Description: Flat or rounded umbels of delicate blue or white flowers, similar in appearance to Queen Anne's lace. Trachymene works well as a filler or as a graceful mass flower.
Vase Life: 7 to 10 days

TRANSVAAL DAISY see ***Gerbera***

TRITELEIA (see Plate 4)

(tri-te-LAY-a)
Family: Amaryllidaceae (daffodil family)
Name Origin: From the Greek words *tri* (three) and *teleios* (perfect), referring to the floral parts, which are in threes.
Species: *laxa*
Common Names: triteleia, brodiaea
Availability: May through November, with peak supplies May through August
Description: Flower heads at a glance are similar to Agapanthus. However, funnel-shaped florets form a smaller umbel and are not as compact and globular as those of Agapanthus; a good filler and accent.
Vase Life: 7 to 10 days

TRITOMA see ***Kniphofia***

TROLLIUS

(TROH-lee-us)
Family: Ranunculaceae (buttercup family)
Species: *Trollius* species
Common Name: globeflower
Availability: April through July
Description: Globe-shaped yellow, orange, or white flowers. Useful in adding mass and accent.
Vase Life: 5 to 7 days

TRUMPET LILY see ***Zantedeschia***
TUBEROSE see ***Polianthes***
TULIP see ***Tulipa***

TULIPA (see Plate 28)

(TEW-li-pa)
Family: Liliaceae (lily family)
Name Origin: From the Turkish word *tulband* (turban).
Species: Many cultivars
Common Name: tulip
Availability: November through May, with peak supplies January through April
Description: Single rounded flowers with colorful sepals and petals; adds mass to mixed spring bouquets. Tulips also work well in oriental and vegetative design styles.
Vase Life: 3 to 7 days; leave stems wrapped in sleeves while hydrating to prevent bending; cut off white portion of lower stem for better water uptake; do not store or arrange with freshly cut daffodils that secrete a sap harmful to tulips.

TURTLEHEAD see ***Chelone***

V

VACCARIA see ***Saponaria***

VANDA (see Plate 19)

(VAN-dah)
Family: Orchidaceae (orchid family)
Species: *coerulea, sanderana, teres, tricolor;* others and hybrids
Common Names: vanda, lei orchid
Availability: Year-round
Description: Sprays with clusters of flat-looking orchids, 1 to 3 inches across. A wide selection of colors with spotted patterns, many of which are fragrant. Many types are used in leis.
Vase Life: 1 to 3 weeks

VERONICA

(ver-RON-ik-ah)
Family: Scrophulariaceae (foxglove family)
Name Origin: Named after St. Veronica.
Species: *spicata, longifolia;* others and cultivars
Common Names: veronica, speedwell
Availability: Year-round, with peak supplies June through September
Description: Upright spikelike racemes made of small purple, blue, pink, or white flowers; useful as a filler; the flower tips curve, adding graceful lines to mixed arrangements.
Vase Life: 5 to 7 days

VERONICA

VIBURNUM

(vy-BUR-num)
Family: Caprifoliaceae (elders and honeysuckles)
Species: *opulus;* others and cultivars
Common Names: snowball, Guelder rose
Availability: January through May
Description: Showy ball-shaped terminal panicles or umbel-like cyme clusters of white or greenish florets on woody stems. These fragrant flowers add mass and texture to large mixed bouquets. Frequently used in large wedding and church decorations.
Vase Life: 5 to 7 days

W

WARATAH see ***Telopea***

WATSONIA

(wot-SONE-ee-ah)
Family: Iridaceae (iris family)

VIBURNUM
snowball

Name Origin: After Sir William Watson (1715–1787), London physician and botanist.
Species: Many species and hybrids
Common Name: watsonia
Availability: March through August
Description: Delicate spikes of florets, similar in appearance to crocosmia; available in whites, pinks, reds, and lavenders.
Vase Life: Long lasting, 7 or more days

WATSONIA

WATTLE see ***Acacia***
WAXFLOWER see ***Chamelaucium***
WILD QUEEN ANNE'S LACE see ***Ammi***
WINDFLOWER see ***Anemone***

Y

YARROW see ***Achillea***
YOUTH AND OLD AGE see ***Zinnia***

Z

ZANTEDESCHIA (see Plate 5)

(zan-te-DES-kee-ah)
Family: Araceae (the aroids)
Name Origin: Named after Italian botanist Francesco Zantedischi (born 1797).
Species: *aethiopica, elliottiana, rehmannii;* others and cultivars
Common Names: calla lily, calla, arum lily, trumpet lily
Availability: Year-round, with peak supplies in spring and summer
Description: A striking white, greenish, yellow, or reddish spathe surrounds a yellow cylindrical spadix. Available in a variety of sizes. Stems are thick and long with heart- or spear-shaped and long-stemmed leaves. Distincitve shape and sleek line provide emphasis.
Vase Life: Varies greatly, from 3 to 14 days

ZINGIBER

(ZIN-ji-ber)
Family: Zingiberaceae (ginger, cardamom, and turmeric)
Species: *zerumbet;* others
Common Name: shampoo ginger
Availability: June to September
Description: Florets appear inside waxy bracts; the entire cluster is similar in appearance to a pinecone. Available in

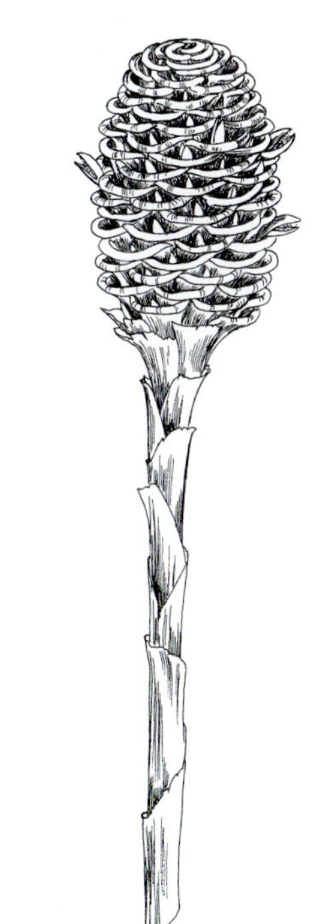

ZINGIBER
shampoo ginger

ZINNIA

pinks, reds, and yellow. Shampoo ginger provides emphasis and combines well with other tropical and exotic flowers.
Vase Life: 7 to 10 days

ZINNIA

(ZIN-ee-ah)
Family: Compositae or Asteraceae (sunflower family)
Name Origin: Named after Johann Gottfried Zinn (1727–1759).
Species: *elegans* cultivars
Common Names: zinnia, youth and old age
Availability: May through October
Description: Daisy-shaped flowers available in a variety of colors, forms, and sizes. These mass flowers are colorful additions to mixed summer arrangements.
Vase Life: 5 to 7 days

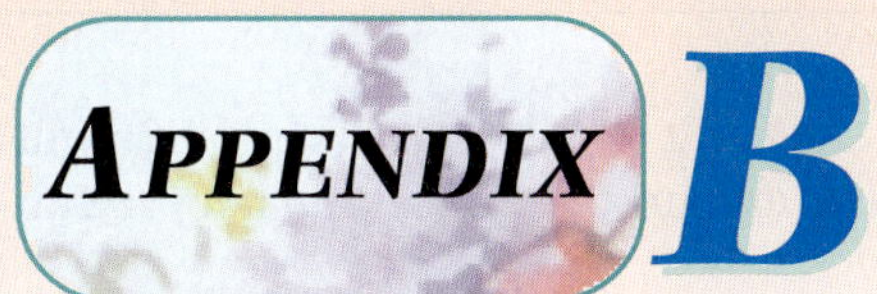

Foliages

The foliage appendix lists information to help you become more familiar with and confident in pronouncing, selecting, and caring for cut foliages. Generic names and common names are listed alphabetically. The name of each ***genus*** is given first, which is also the ***scientific name, botanical name,*** and the ***Latin name,*** and is followed by a suggested pronunciation. The family in which it is placed, the derivation of the name, ***species,*** and ***common names*** are listed for each foliage. Availability and a brief description of each foliage is given along with approximate vase life and specific care and handling procedures. Common names are cross-referenced to the correct scientific name. The color plates, line illustrations, and photographs help ease identification.

Although there is a tremendous variety of foliage to choose from, foliage is too often an afterthought when it comes to gathering the parts to make a flower arrangement. However, it is often the diverse foliage that can set your designs apart from all others, giving them distinction and beauty. Give thought to the foliage you select, its leaf shape, vein pattern, color, texture, and margin. By increasing your design awareness, you also enhance your design skills. For an in-depth look at leaf and foliage characteristics, see Chapter 9.

A

ABIES (see *conifer* illustration)

(AY-beez)
Family: Pinaceae (pine family)
Species: *alba, procera, balsamea;* many others
Common Names: fir, balsam fir; many others
Availability: Commercially, October through January
Description: A large genus of coniferous evergreen trees; stiff branches and aromatic foliage; excellent filler, providing accent with other conifer foliage in winter or Christmas arrangements. Noble fir has bluish, stiff needles with rounded tips that densely cover branched stems. Balsam fir has dark green, rounded needles on branching stems.
Vase Life: Long lasting, from 3 to 4 weeks

ACACIA

(a-KAY-sha)
Family: Leguminosae or Fabaceae (pea family)
Name Origin: From the Greek word *akis* (sharp point), referring to the thorns.
Species: *cultriformis*
Common Names: knifeblade acacia, knife acacia
Availability: Limited supplies year-round
Description: Tall branches with bluish knife-shaped leaves and crowded heads of flowers in terminal racemes; distinctive shape, color, and texture adds line and accent.
Vase Life: 7 to 10 days

ADIANTUM

(a-dee-AN-tum)
Family: Polypodiacea or Adiantaceae (ferns)
Name Origin: From the Greek word *adiantos* (unwetted), referring to the fronds' way of repelling water.
Species: *raddianum;* others and cultivars
Common Name: maidenhair fern
Availability: Year-round
Description: Soft green foliage with shiny black leafstalks; useful as a filler or accent in smaller bouquets and wedding designs.

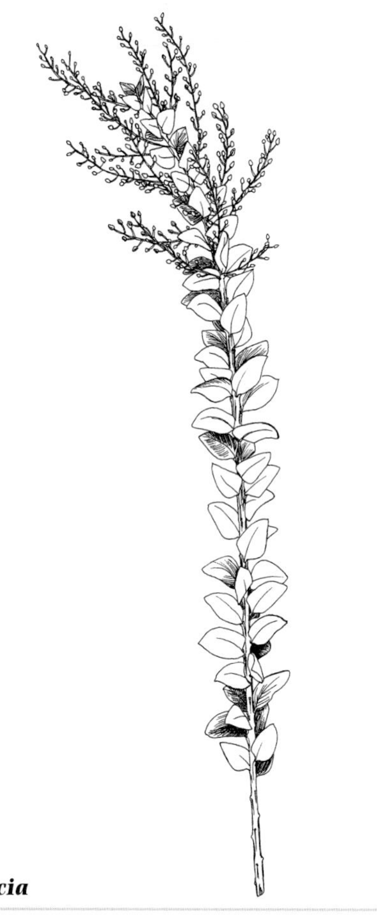

ACACIA
knifeblade acacia

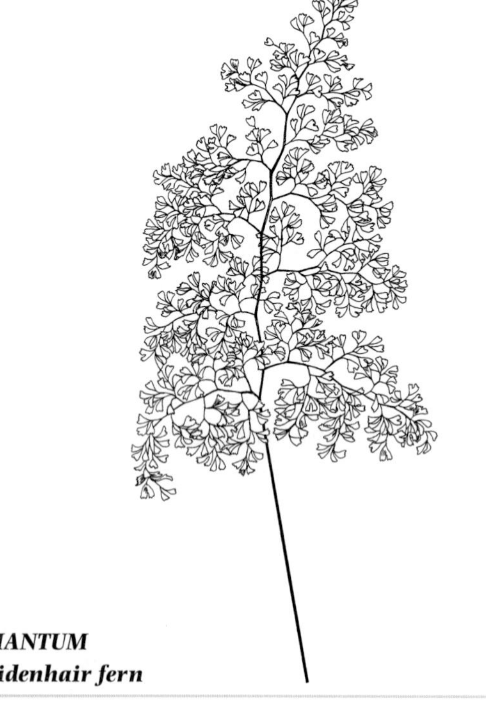

ADIANTUM
maidenhair fern

Vase Life: Short vase life of 2 to 4 days; maidenhair is wilt sensitive; soak in water and mist with water to help regain turgor.

AFRICAN BOXWOOD see ***Myrsine***

ANGEL'S WINGS see ***Caladium***

ANTHURIUM

(an-THEWR-ree-um)
Family: Araceae (the aroids)
Name Origin: From the Greek words *anthos* (flower) and *oura* (tail), referring to the tail-like inflorescence (spadix) of these flowers.
Species: *andraeanum, scherzeranum*
Common Name: flamingo flower foliage
Availability: Year-round, however anthurium foliage is not common, as it is easily damaged while still on the plant.
Description: Heart-shaped like the anthurium flowers. The leaves are dark glossy green and vary in size. Leaves work well with the anthurium flowers and other exotic flowers. Useful as an accent foliage.
Vase Life: 10 to 14 days

ARACHNIODES see ***Rumohra***

ARCTOSTAPHYLOS (see Plate 35)

(ark-tuh-STAF-uh-los)
Family: Ericaceae (heath family)
Name Origin: From the greek words *arctos* (bear) and *staphyle* (bunch of grapes). Manzanita means "little apple" in Spanish.
Species: *manzanita;* others
Common Name: manzanita

ANTHURIUM

ASPARAGUS—common ferns

Availability: Limited supplies year-round
Description: Evergreen shrub, admired for its mahogany-like, smooth, red to purple bark and crooked branches that twist and gnarl attractively.
Vase Life: 7 to 14 days

ASPARAGUS (see Plate 29)

(as-PAIR-a-gus)
Family: Liliaceae (lily family)
Common Names, Species, and Descriptions:

foxtail fern (*A. densiflorus* 'Meyeri'): foxtail fern resembles a green, fluffy foxtail, hence its name; its linear form works well in contemporary designs.

ASPARAGUS
foxtail fern

Ming fern (*A. densiflorus* 'Myriocladus'): (see Plate 29) ming fern is light to dark green filler foliage; the tiny tufts of soft, needlelike leaves appear on branching sprays; excellent foliage for oriental designs; small tufts work well in corsages and boutonnieres.

plumosa fern, lace fern (*A. plumosus*, see Plate 29): plumosa fern ranges from light to dark green, and most types have either a trailing or an upright appearance; this soft, lacy foliage is a delicate linear filler; small pieces work well in corsages and boutonnieres.

sprengeri fern, sprenger fern (*A. densiflorus* 'Sprengeri', see Plate 29): sprengeri fern is generally dark green, but new shoots are often light green; has needlelike leaves densely packed on trailing stems; often will have small greenish berries; useful as a filler and an accent; can create graceful curving lines in large sympathy designs; beautiful accent in bridal bouquets and casket sprays.

string smilax, greenbrier (*A. asparagoides*): string smilax is medium green with small elliptical leaves, closely spaced along a thin stem that grows around a piece of string, forming delicate garlands; commonly used to decorate wedding cakes and banquet tables; useful whenever a soft garland is needed.

tree fern (*A. pyramidalis*): tree fern is medium to dark green and appears in bushy plumes; a soft and airy filler in arrangements; small pieces work well in corsages and boutonnieres.

Availability: Year-round
Vase Life: All are prone to premature drying and shattering; while in the cooler, store in moist bags or in plastic-covered buckets; lasts 3 to 14 days.

ASPARAGUS ASPARAGOIDES
string smilax

ASPIDISTRA (see Plate 30)

(as-pi-DIS-tra)
Family: Liliaceae (lily family)
Name Origin: From the Greek word *aspideon* (a small round shield), referring to the shape of the stigma.
Species: *elatior*
Common Names: cast iron plant, barroom plant
Availability: Year-round, with limited quantities in the winter months
Description: Shiny dark green or variegated color stripes and color blotches. These leaves look similar to *Cordyline terminalis* (ti leaves) but have a softer texture and do not have a prominent center midrib. Useful as a background foliage; common for large designs.
Vase Life: Long lasting 3 to 4 weeks

AUSTRALIAN LAUREL see ***Pittosporum***

BUXUS
boxwood
BUXUS
Oregonia
XEROPHYLLUM
bear grass
ASPIDISTRA
cast iron plant

ASPARAGUS PYRAMIDALIS
tree fern

B

BAKER FERN see ***Rumohra***
BALSAM see ***Abies***
BARROOM PLANT see ***Aspidistra***
BEAR GRASS see ***Xerophyllum***
BELLA PALM see ***Chamaedorea***
BIRD OF PARADISE FOLIAGE see ***Strelitzia***
BOSTON FERN see ***Nephrolepis***
BOXWOOD see ***Buxus***
BRAKE FERN see ***Nephrolepis***
BROOM see ***Cytisus***
BUTCHER'S BROOM see ***Ruscus***

BUXUS (see Plate 30)
(BUCK-sus)
Family: Buxacea (box family)
Species: *sempervirens;* others and cultivars
Common Names: boxwood, box (*B. semperivirens*); Oregonia *(B. species)*
Availability: Year-round
Description: Boxwood has small, oval, dark green, glossy leaves that densely cover woody stems. Oregonia is similar but with a variegated coloring. Useful as a filler and for providing line.
Vase Life: Long lasting, 1 to 2 weeks

C

CABBAGE PALM see ***Sabal***

CALADIUM (see Plate 31)
(ka-LAY-dee-um)
Family: Araceae (the aroids)
Name Origin: From the native name, *kaladi.*
Species: *bicolor* and hybrids
Common Names: angel's wings, elephant's ear, heart of Jesus, mother-in-law plant
Availability: Year-round
Description: Caladium plants are grown for their interesting foliage. They have large, heart- shaped leaves with contrasting, marbled color patterns and venation. Wide range of colors including greens, white, cream, pinks, and reds; provides emphasis and accent.
Vase Life: 3 to 5 days

CALATHEA (see Plate 31)
(ka-lah-THEE-a)
Family: Marantacea (arrowroot)
Name Origin: From the Greek word *kalathos* (basket), referring to the flowers being clustered as if in baskets.
Species: *zebrina;* others
Common Names: zebra plant, peacock leaf
Availability: Year-round
Description: All species have leaves with interesting colors and venation patterns; works well as an accent foliage, and can easily create or enhance a focal point. Calathea harmonizes well with exotic flowers.
Vase Life: 1 to 3 weeks

CAMELLIA (see Plate 31)
(ka-MEL-lee-a)
Family: Theacea (tea, camellias, and franklinia)
Name Origin: Named after George Joseph Kamel (1661–1706), who studied the Philippines flora.
Species: *japonica;* others
Common Name: camellia
Availability: Year-round
Description: Handsome, glossy green leaves on long woody branches. An excellent background foliage for large arrangements. Smaller branches are useful as accents or fillers. Individual leaves work well in corsages and boutonnieres.
Vase Life: 5 to 7 days

CAST IRON PLANT see ***Aspidistra***
CEDAR see ***Cedrus***

CALATHEA
zebra plant
CAMELLIA
CODIAEUM
croton
CALADIUM
angel's wings
EUONYMUS

CEDRUS (see *conifer* illustration)

(SEE-drus)
Family: Pinaceae (pine family)
Species: *atlantica, deodara;* others and cultivars
Common Name: cedar, atlas cedar, deodar cedar; others
Availability: Year-round
Description: Cedar is flat and lacy with stiff needles in clusters; provides a soft, sweeping line; smaller pieces work well as fillers.
Vase Life: Long lasting, 3 to 4 weeks

CHAMAEDOREA (see *palm* illustration)

(ka-mee-DOR-ee-ah)
Family: Palmae or Arecaceae (palm family)
Name Origin: From the Greek words *chamai* (on the ground) and *dorea* (gift), referring to the fruits that are within reach, unlike most palms.
Species: *elegans, oblongata;* others
Common Names: giant palm, parlor palm *(C. elegans)*; bella palm, narrow palm *(C. elegans* 'Bella'); others
Availability: Year-round

CONIFERS

Description: Palms are medium to dark green. Giant palm has sessile, elongated 1-inch leaves. Narrow palm leaves are ½ inch wide. Both are useful in large arrangements. Excellent background foliage. Can be trimmed for cleaner geometric or abstract patterns.
Vase Life: 5 to 7 days

CLUB MOSS see ***Lycopodium***

CODIAEUM (see Plate 31)

(koh-die-EE-um)
Family: Euphorbiaceae (spurge family)
Species: *variegatum* cultivars
Common Name: croton
Availability: Year-round
Description: Colorful, variegated, bold patterns on leaves that vary in size and shape. Wide range of colors including red, orange, pink, green, yellow, and white. Croton leaves can create a focal point or add accent.
Vase Life: 1 week

CONIFER see ***Abies, Cedrus, Juniperus, Pinus***

CORDYLINE (see Plate 37)

(kor-di-LIE-nee)
Family: Agavaceae (sisal hemp, pulque, and dragon tree)
Name Origin: From the Greek word *kordyle* (club), referring to the large and fleshy roots of some species.
Species: *terminalis* (formerly *Dracaena terminalis);* cultivars
Common Names: ti (tea) leaf, ti, good luck plant, tree of kings
Availability: Year-round
Description: Long glossy leaves are emerald green or darker green with a reddish margin. Leaf lengths and widths vary. Useful as background foliage, especially with tropical and exotic flowers. Leaves can be curled or cut into many shapes.
Vase Life: 1 to 2 weeks

CROTON see ***Codiaeum***

CYCAS

(SIE-kas)
Family: Cycadacea (cycas family)
Name Origin: From the Greek name for a palm.
Species: *revoluta*
Common Names: cycas palm, sago palm
Availability: Year-round
Description: Stiff, short, dark green leaves on straight stems, about 20 inches long; leaflets have margins bent downward with sharp tips. Useful with tropical and large arrangements.
Vase Life: Long lasting, 1 to 3 weeks

DRACAENA
EUCALYPTUS
spiral eucalyptus
EUCALYPTUS
seeded eucalyptus
DIFFENBACHIA
dumb cane
EUPHORBIA
snow on the Mountain

CYPERUS PAPYRUS
papyrus

CYPERUS
(sie-PEE-rus *or* SY-per-us)
Family: Cyperacea (reeds and sedges)
Name Origin: From the Greek word for a sedge.
Species: *alternifolius, papyrus*
Common Names: papyrus, bulrush, paper plant (*C. papyrus*); umbrella palm (see *palms* illustration), palm crown (*C. alternifolius*)
Availability: Year-round
Description: Cyperus (umbrella palm) is medium green and quite showy. Its radiating umbrella appearance adds a striking form to contemporary design styles. Papyrus is also medium green but has a moplike tuft of short, stiff, grasslike leaves at the top of each stem. Unusual foliage that adds interest, texture, and form. Looks best combined with unusual and exotic flowers or in contemporary design.
Vase Life: Long lasting, 2 to 4 weeks

CYTISUS (see Plate 37)
(SIT-is-us *or* si-TIS-sus)
Family: Leguminosae or Fabaceae (pea family)
Name Origin: From the Greek word *kytisos*, the name for these and similar shrubs.
Species: *scoparius* cultivars
Common Name: Scotch broom, broom
Availability: August through April
Description: Long, needlelike, dark green leaves that appear stiff on woody stems; the branches can be shaped into curves by gently using the warmth of your hands to shape, and by blowing hot air on the Scotch broom to speed the curving process; it is an ideal foliage for setting the shape of crescent and Hogarth designs. Another plant called Ironwood or Australian pine, is often mistaken for Scotch broom.
Vase Life: Long lasting, up to 3 weeks

D

DIFFENBACHIA (see Plate 32)
(deef-en-BAHK-ee-ah)
Family: Araceae (the aroids)
Name Origin: Named afer J. F. Dieffenbach (1790–1863).
Species: *imperialis*; others and cultivars
Common Name: dumb cane
Availability: Limited; generally leaves are cut from a potted plant.
Description: Handsome leaves have variegated color patterns, spotted and feathered with white, cream, and yellow markings. Diffenbachia adds emphasis and accent to designs.
Vase Life: Varies with species; 1 to 2 weeks

DRACAENA (see Plate 32)
(dra-SEE-nah)
Family: Agavaceae (sisal hemp, pulque, and dragon tree)
Name Origin: From the Greek word *drakcina* (dragon).
Species: *cincta, concinna;* others and cultivars
Common Name: dracaena; others. Many properly belong to the closely related genus *Cordyline*, but are sold and grown as dracaena.
Availability: Year-round
Description: Long slender leaves, from dark to light green with variegated striping patterns; adds line and accent to floral designs.
Vase Life: Varies with species; 3 to 4 weeks

DRYOPTERIS see ***Rumohra***
DUMB CANE see ***Diffenbachia***

VACCINIUM
huckleberry
GALAX
galax leaf
ILEX
holly
PHORMIUM
New Zealand flax
PANDANUS
hala leaf
ULEX
gorse

E

ELEPHANT'S EAR see ***Caladium***
ELK GRASS see ***Xerophyllum***

EQUISETUM
(ek-wi-SEE-tum)
Family: Equisetaceae (equisetum)
Name Origin: From the Latin words *equus* (horse) and *seta* (bristle).
Species: *hyemale*
Common Names: horsetail, snake grass, scouring rush
Availability: Year-round
Description: Hollow, jointed dark green stems; leaves are mere scales at the joints. Spores are borne in conelike spikes at the top of each stem. Useful for adding line; easily manipulated into abstract linear shapes.
Vase Life: Long lasting, 1 to 2 weeks

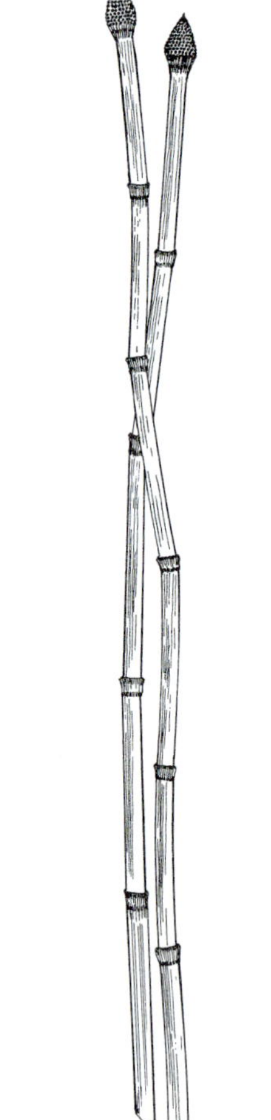

EQUISETUM
snake grass

EUCALYPTUS (see Plate 32)
(ew-ka-LIP-tus)
Family: Myrtaceae (myrtles, eucalyptus, and cloves)
Name Origin: From the Greek words *eu* (well) and *kalypto* (to cover), referring to how the calyx forms a lid over the flowers in bud.
Species: *cinerea, gunnii, nicholii, perriniana, polyanthemos, pulverulenta, tetragona, torquata;* others
Common Names: seeded eucalyptus and spiral eucalyptus (see Plate 32), baby blue eucalyptus, silver dollar eucalyptus, willow eucalyptus, silver spoon eucalyptus, eucalyptus pods (see Plate 10); others
Availability: Year-round
Description: Most eucalyptus is bluish green to silver-gray; leaves vary in shape; some are rounded and sessile along tall stems that are useful for establishing line; other types have oval or slender, elongated leaves; most have a medicinal fragrance; removing lower leaves produces a sticky residue called menthol; remove by scrubbing hands with soap that contains lanolin.
Vase Life: Long lasting, 1 to 4 weeks

EUCALYPTUS POLYANTHEMOS
silver dollar eucalyptus

EUCALYPTUS NICHOLII
willow eucalyptus

EUONYMUS (see Plate 31)
(ew-ON-i-mus)
Family: Celastraceae (spindle tree family)
Name Origin: From the Latin name for these deciduous and evergreen trees and shrubs.
Species: *japonica, fortunei;* others
Common Name: euonymus
Availability: Year-round
Description: Evergreen foliage is glossy green and variegated. The leaves are oval and densely grouped on woody stems. Useful as a background foliage; shorter stems work well as an accent and filler.
Vase Life: Long lasting, 2 to 3 weeks

HEDERA
ivy
RUMOHRA
leatherleaf fern
MAGNOLIA
MAHONIA
Oregon grape

EUPHORBIA (see Plate 32)

(ew-FOR-bee-a)
Family: Euphorbiaceae (spurge family)
Name Origin: Named after Euphorbus, the physician to the King of Mauritania, Juba.
Species: *marginata*
Common Name: snow on the mountain, ghostweed
Availability: Year-round
Description: White, yellow, and light green leaves with contrasting white margin; it exudes a white milky sap when cut that can irritate sensitive skin and cause severe burning or dermatitis. Beautiful accents in foliage arrangements and contemporary designs.
Vase Life: 5 to 7 days; seal stems for increased vase life; change water often.

F

FATSIA

(FATS-see-a)
Family: Araliaceae (ivies and ginseng)
Species: japonia cultivars
Common Name: Japanese aralia
Availability: Mostly summer months
Description: Large, glossy, dark green leaves, deeply lobed like the fingers of a hand. Beautiful additions in foliage arrangements. Useful as background or filler in large arrangements.
Vase Life: Long lasting, 1 to 2 weeks

FATSIA
Japanese aralia

FERN see ***Nephrolepis***
FIR see ***Abies***
FLAT FERN see ***Nephrolepsis***
FLAX see ***Phormium***
FOXTAIL FERN see ***Asparagus***
FURZE see ***Ulex***

G

GALAX (see Plate 33)

(GAY-lax)
Family: Diapensiaceae (shortia and galax)
Name Origin: From the Greek word *gala* (milk), referring to the white flowers.
Species: *urceolata*
Common Name: galax leaf
Availability: Year-round; limited supplies in May and June
Description: Single, heart-shaped or rounded leaves on short stems. Generally dark green; however, in the autumn and winter months galax is reddish green. Galax is useful as a background foliage for sympathy pieces and wreaths. Works well as an accent foliage in contemporary and oriental sytles.
Vase Life: 1 to 2 weeks

GAULTHERIA (see Plate 37)

(gawl-THER-ee-a *or* gawl-the-REE-ah)
Family: Ericaceae (heath family)
Name Origin: Named after a Canadian botanist and physician, Dr. Gaulthier (1708–1758).
Species: *shallon*
Common Names: salal (sa-LAL), lemonleaf, shallon
Availability: Year-round, with lesser quantities in July
Description: Large, medium to dark green leaves on long woody branches. Good background foliage in large arrangements; smaller tips work well as fillers. Individual leaves are useful in corsage and boutonnieres.
Vase Life: Long lasting, 3 weeks

GIANT PALM see ***Chamaedorea***
GOOD LUCK PLANT see ***Cordyline***
GORSE see ***Ulex***
GREENBRIAR see ***Asparagus***
GROUND PINE see ***Lycopodium***

MYRTUS
myrtle

LYCOPODIUM
club moss

ARCTOSTAPHYLOS
manzanita

H

HALA see ***Pandanus***
HEART OF JESUS see ***Caladium***

HEDERA (see Plate 34)
(HED-er-ah)
Family: Araliaceae (ivies and ginseng)
Name Origin: From the Latin name for these evergreen climbers.
Species: *helix, canariensis, colchica;* cultivars
Common Names: ivy, English ivy *(H. helix);* Algerian ivy *(H. canariensis);* Persian ivy *(H. colchica)*
Availability: Year-round
Description: Green or variegated leaves, spaced on flexible stems. Ivy has a graceful trailing quality. Useful as a hanging and curving line foliage, and as a filler foliage.
Vase Life: 5 days or longer

HOLLAND RUSCUS see ***Ruscus***
HOLLY see ***Ilex***
HORSETAIL see ***Equisetum***
HUCKLEBERRY see ***Vaccinium***

I

ILEX (see Plate 33)
(EYE-leks)
Family: Aquifoliaceae (holly family)
Species: *aquifolium;* others and cultivars
Common Name: holly
Availability: Commercially, November and December
Description: Dark green or variegated stiff, spiky leaves on woody stems with red berries. Useful as a winter or holiday accent and filler. Spiky leaf margins add interesting form and texture to designs.
Vase Life: Long lasting, 1 to 3 weeks

INDIAN BASKET GRASS see ***Xerophyllum***
ITALIAN RUSCUS see ***Ruscus***
IVY see ***Hedera***

J

JAPANESE ARALIA see ***Fatsia***
JUNIPER see ***Juniperus***

JUNIPERUS (see *conifer* illustration)
(joo-NIP-er-us)
Family: Cupressaceae (juniper family)
Name Origin: From the Latin name for these evergreen conifers.
Species: *communis;* others and cultivars
Common Name: juniper
Availability: October through December
Description: Medium to light green needles on spreading branches with powdery blue berries. Generally used as a filler foliage in winter and Christmas arrangements and decorations; texture of foliage and berries provides accent.
Vase Life: Long lasting, 3 to 4 weeks

K

KNIFEBLADE ACACIA see ***Acacia***

L

LACE FERN see ***Asparagus***
LEATHERLEAF FERN see ***Rumohra***
LEMONLEAF see ***Gaultheria***

LYCOPODIUM (see Plate 35)
(lie-koh-POH-dee-um)
Family: Lycopodiaceae (lycopodium)
Species: *complanatum, obscurum;* others
Common Names: club moss, ground pine
Availability: Year-round
Description: Bright green scalelike needles on short, stiff, forking branches. The stems are 15 to 20 inches long. Useful in establishing line. Also useful as a filler in contemporary designs.
Vase Life: 7 to 10 days

M

MAIDENHAIR FERN see ***Adiantum***

MAGNOLIA (see Plate 34)
(mag-NOL-ee-a)
Family: Magnoliaceae (magnolias and tulip tree)
Species: *grandiflora;* others
Common Name: magnolia
Availability: October through March
Description: Large, glossy, dark green leaves with a leathery texture on woody stems. Useful in large arrangements.
Vase Life: About 5 days

PODOCARPUS
tropic yew
variegated pitt
pitt
PITTOSPORUM
RUSCUS
butcher's broom
RUSCUS
Italian ruscus

MAHONIA (see Plate 34)

(ma-HON-ee-ah)
Family: Berberidaceae (barberry, sacred bamboo, and may apple)
Name Origin: Named after American horticulturist Bernart McMahon (died 1816).
Species: *aquifolium;* others
Common Names: mahonia, Oregon grape
Availability: Year-round
Description: Evergreen shrubs with glossy, spiny leaves. Useful as a filler in larger arrangements or as a background foliage.
Vase Life: 5 to 10 days

MANZANITA see *Arctostaphylos*

MARANTA

(ma-RAN-tah)
Family: Marantaceae (arrowroot)
Name Origin: Named after Venetian botanist Bartolommeo Maranti (16th century).
Species: *leuconeura* cultivars
Common Name: prayer plant
Availability: Limited; generally leaves are cut from potted plants.
Description: Light and dark green with variegated color and vein patterns. Useful in adding emphasis and accent.
Vase Life: 3 to 5 days

MELALEUCA

MELALEUCA

(me-la-LOO-ka)
Family: Myrtaceae (myrtle family)
Name Origin: From the Greek words *melas* (black) and *leukos* (white), referring to the black trunk and white shoots of many species.
Species: *neosphila, elliptica;* others
Common Name: melaleuca
Availability: Year-round
Description: Evergreen, woody shrubs, often mistaken for mini myrtle.
Vase Life: Long lasting, 1 to 3 weeks

MING FERN see *Asparagus*

MOCK ORANGE see *Pittosporum*

MONSTERA
Swiss cheese plant

MONSTERA

(mon-STER-ah)
Family: Araceae (the aroids)
Name Origin: Possibly the name derived from the monstrous appearance and size of the leaves.
Species: *deliciosa*
Common Names: monstera, Swiss cheese plant
Availability: Year-round
Description: Large, leathery, glossy, dark green leaves; irregularly cut and perforated with interesting margins and negative spaces, thus the name Swiss cheese plant. Useful foliage with exotic and tropical flowers.
Vase Life: Long lasting, 2 to 4 weeks

MYRSINE

MYRSINE
African boxwood

(MUR-si-nay)
Family: Myrsinaceae (myrsine)
Name Origin: Greek name for myrtle.
Species: *africana*
Common Name: African boxwood
Availability: Year-round
Description: Evergreen woody shrub, often mistaken for myrtle and Buxus species; has rounded, fleshy leaves; tall, like myrtle; dries well.
Vase Life: Long lasting, 1 to 3 weeks

GAULTHERIA
salal

CYTISUS
Scotch broom

CORDYLINE
ti leaves

MYRTLE see ***Myrtus***

MYRTUS
(see Plate 35)
(MUR-tus)
Family: Myrtaceae (myrtle family)
Species: *communis;* others and cultivars
Common Name: myrtle
Availability: October through March
Description: There are several types of myrtle that are used commercially: common myrtle grows up to 48 inches tall with large, glossy, dark green leaves; mini myrtle has lighter-colored and smaller, more compact leaves on shorter stems; variegated myrtle has color patterns with cream and yellow margins. Useful as a line foliage in setting the framework of a design; smaller pieces are useful as fillers. Myrtle has a lemon fragrance.
Vase Life: Long lasting, 1 to 2 weeks

MYRTUS COMMUNIS
mini myrtle

MYRTUS COMMUNIS
variegated myrtle

N

NARROW PALM see ***Chamaedorea***

NEPHROLEPIS
(nef-row-LEP-is)
Family: Polypodiaceae (ferns)
Species: *exaltata, cordifolia;* cultivars
Common Names: Boston fern (*N. exaltata* 'Bostoniensis'); Oregon fern, brake fern, flat fern, sword fern (*N. cordifolia*)
Availability: Year-round
Description: Boston fern has short, light green leaves alongside a short stem usually no longer than 12 inches; an attractive filler and useful in contemporary designs. Oregon fern is much longer and wider than Boston fern. It can easily set the framework of larger designs and is useful in covering mechanics.
Vase Life: 2 to 5 days

NEW ZEALAND FLAX see ***Phormium***

NOBLE FIR see ***Abies***

O

OREGON FERN see ***Nephrolepis***
OREGON GRAPE see ***Mahonia***
OREGONIA see ***Buxus***

P

PALM see ***Chamaedorea***
PALM CROWN see ***Cyperus***
PALMETTO PALM see ***Sabal***
PALM FAN see ***Sabal***

PANDANUS (see Plate 33)
(PAN-da-nus *or* pan-DAN-us)
Family: Pandanaceae (screw pines)
Species: *odaratissimus, utilis;* others
Common Names: screw pine, hala leaf
Availability: Year-round

NEPHROLEPIS

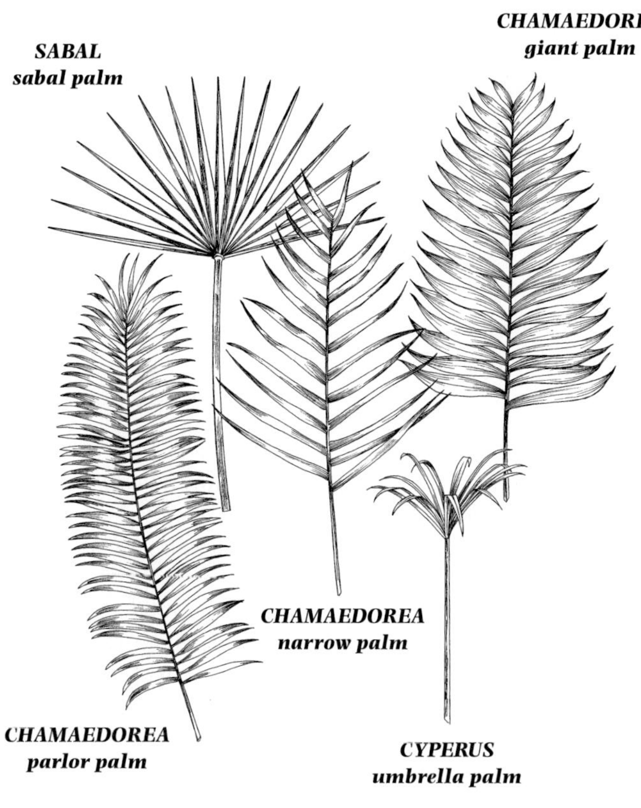

PALMS

Description: Sword-shaped leaves with yellow striping, 24 to 36 inches long. Pandanus, like New Zealand flax (*Phormium*), makes a dramatic vertical statement. It can be curled and bent into abstract, geometric patterns.
Vase Life: Long lasting; 2 to 3 weeks

PAPYRUS see ***Cyperus***
PARLOR PALM see ***Chamaedorea***
PEACOCK LEAF see ***Calathea***

PHORMIUM (see Plate 33)
(FOR-mee-um)
Family: Agavaceae (sisal hemp, pulque, and dragon tree)
Name Origin: From the Greek word *phormion* (mat), referring to the fiber that is produced from the leaves of *P. tenax.*
Species: *tenax* cultivars
Common Names: flax, New Zealand flax
Availability: Year-round
Description: Long, narrow leaves with variegated green and white and reddish purple patterns. Useful in adding line; may be manipulated into abstract linear patterns. Harmonizes well with tropical flowers.
Vase Life: 5 to 14 days

PINE see ***Pinus***

PINUS (see *conifer* illustration)
(PIE-nus)
Family: Pinaceae (pine family)
Species: *strobus;* many others and cultivars
Common Names: pine; white pine; others
Availability: Commercially, October through February
Description: Long flexible needles on branching stems. Sizes and types vary with species.Useful in setting the skeleton of a design; smaller pieces work well as accents and fillers in winter and Christmas arrangements and decorations.
Vase Life: Long lasting, 3 to 4 weeks

PITTOSPORUM (see Plate 36)
(pi-TOS-poh-rum *or* pi-toe-SPORE-um)
Family: Pittosporaceae (parchment-bark)
Name Origin: From the Greek words *pitta* (pitch) and *sporum* (seed), referring to the sticky seeds.
Species: *tobira;* others
Common Names: pittosporum, pitt, Australian laurel, mock orange
Availability: Year-round
Description: Dark to light green, thick, leathery leaves including variegated types with cream-colored margins. Pittosporum can easily set the skeleton or background of a design; useful as a filler foliage.
Vase Life: Long lasting, 1 to 2 weeks

PLUMOSA FERN see ***Asparagus***

PODOCARPUS (see Plate 36)
(poh-doh-KAR-pus)
Family: Podocarpaceae (podocarpus)
Species: *macrophyllus;* others
Common Name: podocarpus, tropical yew
Availability: Year-round
Description: Dark green and conifer-like with long, flexible, flat needles on tall branches. Useful in setting the framework of a design. Shorter branches are useful as accents or fillers.
Vase Life: Long lasting, 1 to 3 weeks

POLYSTICHUM
(po-LI-sti-kum)
Family: Polypodiaceae (ferns)
Name Origin: From the Greek words *poly* (many) and *stichos* (row), referring to the arrangement of the spores on the underside of the fronds.
Species: *munitum;* others
Common Names: western sword fern, shield fern

Availability: Year-round, with lesser quantities in the summer months
Description: Western sword fern is similar to *Nephrolepis* but has longer and wider leaves on stems that are 20 to 24 inches long.
Vase Life: 10 to 14 days

PRAYER PLANT see ***Maranta***

R

RHAMNUS
(RAM-nus)
Family: Rhamnaceae (buckthorn and jujube)
Name Origin: The Greek name of a shrub.
Species: *alaternus;* others
Common Name: variegated buckthorn
Availability: Year-round
Description: Often confused with variegated myrtle or variegated boxwood; has tall woody branching stems.
Vase Life: Long lasting, 1 to 3 weeks

RHAMNUS ALATERNUS variegated buckthorn

RUMOHRA (see Plate 34)
(roo-MOH-rah)
Family: Polypodiaceae (ferns)
Species: *adiantiformis;* other genera of the *Polypodiaceae* family *(Dryopteris, Arachniodes)* are often named leather fern, causing some confusion.
Common Names: leather fern, leatherleaf, baker fern
Availability: Year-round
Description: Dark green, triangular-shaped fronds, coarsely toothed. A popular florist foliage; works well in setting the background shape and in covering mechanics. Smaller pieces work well as fillers and in corsages and boutonnieres.
Vase Life: Long lasting, 1 to 3 weeks

RUSCUS (see Plate 36)
(RUS-kus)
Family: Liliaceae (lily family)
Species: *aculeatus, hypoglossum*
Common Names: butcher's broom, Holland ruscus *(R. aculeatus);* Italian ruscus, smilax ruscus *(R. hypoglossum)*
Availability: Year-round
Description: Butcher's broom has broad green leaves on stems 10 to 20 inches long. Adds an interesting line; also useful as a filler. Italian ruscus has branching stems each with small, green leaves resembling string smilax. Provides sweeping curves in designs or can be used as a filler.
Vase Life: 7 to 10 days

S

SABAL (see *palm* illustration)
(SAY-bal)
Family: Palmae or Arecaceae (palm family)
Species: *palmetto*
Common Names: sabal, palmetto palm, palm fan, cabbage palm
Availability: Year-round
Description: Palmate leaves, deeply divided and forming a fan about 20 inches wide. Palmetto works well as a background foliage in large designs. May be trimmed for interesting shapes.
Vase Life: Long lasting, 1 to 2 weeks

SAGO PALM see ***Cycas***
SALAL see ***Gaultheria***
SCOTCH BROOM see ***Cytisus***
SCOURING RUSH see ***Equisetum***
SCREW PINE see ***Pandanus***
SHALLON see ***Gaultheria***
SHIELD FERN see ***Polystichum***
SILVER DOLLAR see ***Eucalyptus***
SMILAX RUSCUS see ***Ruscus***
SNAKE GRASS see ***Equisetum***
SNOW ON THE MOUNTAIN see ***Euphorbia***
SPRENGERI FERN see ***Asparagus***
SPURGE see ***Euphorbia***

STRELITZIA
(stre-LITZ-ee-a)
Family: Strelitziaceae (bird of paradise)
Species: *reginae*
Common Name: bird of paradise
Availability: Year-round
Description: Oval-shaped, light to dark green leaves; up to 20 inches long on sturdy straight stems. Ideal foliage for tall, tropical arrangements.
Vase Life: Long lasting, 1 to 2 weeks

STRELITZIA
bird of paradise

STRING SMILAX see ***Asparagus***
SWISS CHEESE PLANT see ***Monstera***

T

TI LEAF see ***Cordyline***
TREE FERN see ***Asparagus***
TREE OF KINGS see ***Cordyline***
TROPICAL YEW see ***Podocarpus***

U

ULEX (see Plate 33)

(EW-leks)
Family: Leguminosae or Fabaceae (pea family)
Species: *europaeus*
Common Names: gorse, furze, whin
Availability: June through April
Description: Rigid, dark green, spiny branches. Excellent for adding line; useful as a background element. Provides accent in contemporary designs.
Vase Life: 5 to 7 days; sensitive to water stress

UMBRELLA PALM see ***Cyperus***

V

VACCINIUM (see Plate 33)

(vak-SIN-ee-um)
Family: Ericaceae (heath family)
Species: *ovatum*
Common Names: huckleberry, huck
Availability: Year-round, with lesser quantities in July
Description: Small ovate leaves densely cover branching stems. Useful as a background foliage in large arrangements. Smaller branches work well as fillers.
Vase Life: Long lasting, 1 to 2 weeks

VARIEGATED BUCKTHORN see ***Rhamnus***

W

WESTERN SWORD FERN see ***Polystichum***
WHIN see ***Ulex***
WHITE PINE see ***Pinus***

X

XEROPHYLLUM (see Plate 30)

(zer-oh-FIL-um)
Family: Liliaceae (lily family)
Species: *tenax*
Common Names: bear grass, elk grass, Indian basket grass
Availability: Year-round
Description: Long, thin, curving leaves, about $\frac{1}{4}$ inch wide and up to 48 inches long. Bear grass adds a soft curving line to designs, particularly in otherwise stiff arrangements. Good foliage to add to waterfall-style bouquets.
Vase Life: Long lasting, 1 to 3 weeks

Z

ZEBRA PLANT see ***Calathea***

Glossary

abscission. The dropping off of leaves and flowers, which rapidly increases in the presence of ethylene.

abstract design. Free-form design in which plant material and nonfloral items are used in a nonnaturalistic way; emphasis is placed on color, texture, and form.

abstract Japanese design/style. A style of Japanese design that emphasizes forms and textures, and often uses nonplant materials with flowers; also called *freestyle* or *jiyu-bana,* this style was introduced following the end of WWII; these designs express unique individualism instead of depicting traditional nature scenes.

accent. A subordinate element that enhances the primary structure of a composition. Also, to call attention to a particular location in a floral design; used to create emphasis.

accent cluster. A small floral design added to a larger composition or area, particularly in large funeral pieces.

accessories. Any material or object, other than plant material, such as candles, bows, novelties, and plush toys, added to enhance the theme of a design. Also may refer to plant materials other than flowers and foliage, such as pinecones, sticks, bamboo, and berries.

achromatic. Without color. Any gradation of white, gray, or black.

acidic. When referring to the pH level of water for cut flowers, the lower the pH level, the more hydrogen ions it contains (acidic). The ideal preservative solution for cut flowers is 3.0 to 4.5; this water has better cohesive qualities and moves more readily into and through stems; microbial growth is limited in solutions that are acidic.

acidifier. To make or become acidic; commercial floral preservatives contain acidifiers for reducing the pH level of water, maximizing water uptake; effective acidifiers in preservatives include citric acid and aluminum sulfate.

actual line. Real lines that lead the eye throughout a composition; curving branches and flower stems are real lines.

advanced design. Refers to floral styles that are complex, requiring specialized skills, mechanics, and techniques.

advertising. Paid promotional activities. Supplying information to the public to induce people to buy a product.

advertorial. Captivating printed advertisement designed to look and read like a magazine article.

air-drying. A simple method of drying fresh flowers and other plant material. Hanging upside down, drying upright, or standing in water and allowing moisture in the plant to dissipate.

air embolism. The blockage by air emboli or bubbles that inhibits the uptake of water and nutrients.

aisleabra. Candelabra that are attached to the ends of church pews or chairs on the main aisle for wedding ceremonies.

aisle runner. A white carpet running the length of the center aisle in a church for wedding ceremonies.

alkaline. Having a pH greater than 7. The alkalinity of water indicates the water's capacity for pH change with the addition of preservatives and other chemicals.

allied florists' association. A group of growers, wholesalers, and retailers in an area, contributing a percentage of their gross income for the advertising and promotion of flowers.

all-sided design. A floral arrangement designed to be viewed from all sides.

altar decorations. General term for floral arrangements that are placed on and around the raised platform area (altar) of the church; these designs are often large and symmetrical and help to visually frame the bride and groom during the ceremony. See also *candelabra.*

alternate complement. A four-color scheme combining a triad and the direct complement of one of the colors.

aluminum sulfate. A white crystalline substance that is high in salts used to purify and acidify water for cut flowers.

American Academy of Floriculture (AAF). A division of the Society of American Florists dedicated to service in the floral industry and excellence in leadership.

American Floral Marketing Council (AFMC). A branch of the Society of American Florists that promotes and advertises the sale of flowers, plants, and floral products throughout the year for nonoccasion days.

American Institute of Floral Designers (AIFD). The premier organization of floral designers established to enact and maintain standards of excellence in the field.

American style. A general description for floral arrangements having a combination of line and mass.

analogous color scheme. A color scheme utilizing several adjacent colors on the color wheel, such as yellow, yellow-orange, and orange.

Analysis of Beauty, The. Title of a work published in 1753. William Hogarth theorized that all beauty was based on the serpentine S-line.

anatomy. In botany, the area that deals with the internal structures of organisms.

anchor pin. Round, plastic holder with four upright prongs. When glued inside the bottom of a container, the prongs hold floral foam in place.

anchor tape. Waterproof tape used primarily to hold floral foam in place.

androecium. The collective name for all the stamens in a flower.

angling. Technique used to give one-sided designs a fuller, more balanced look by exaggerating the stem positioning of flowers in a design; tallest background stems are directed to the back while lowest foreground stems are tilted downward in front of the container.

anther. The pollen-bearing portion of a stamen. In some flowers, such as lilies, the anthers should generally be removed prior to arrangement and delivery because the pollen of the anthers stains clothing.

antitranspirant. A liquid spray or dip that protects the surface of a flower or foliage and minimizes water loss or transpiration.

arm bouquet. A tied cluster of flowers carried across the forearm for weddings or as a presentation bouquet.

Art Deco. A term derived from the 1925 Paris exhibition *Les Expositions des Arts Decoratifs*; known during the 1920s and 1930s as modernistic. General term describing architecture, furniture, decorative arts, and floral arrangements having strong, streamlined geometric forms, lines, and patterns including zigzags, pyramids, and sunburst motifs. Floral designs feature geometrically bold containers and form flowers and foliage. Thearrangements are formal linear or high style in appearance.

Art Nouveau. A modern stylistic movement based on the flowing lines of nature that flourished principally in Europe and the United States around 1890–1910. These floral arrangements are often characterized by lavish cascading asymmetrical waterfall-style designs.

asymmetrical balance. Having unequal visual weight on either side of an imaginary center vertical axis. Often referred to as optical (connected with the sense of sight) or occult (hidden and concealed) balance.

asymmetrical design. A floral arrangement having unequal visual weight on either side of an imaginary center vertical axis; appears more natural and is often associated with informal occasions and settings.

attar. The fragrant oil in flowers stored in the epidermal cells of petals or petal substitutes.

auction. A centralized facility, common in Holland, where numerous growers bring their floral products to sell through open-market bidding by distributors, exporters, wholesalers, and retailers.

axil. The upper angle formed between a leaf and a stem.

balance. A design principle; the placement of flowers, foliage, and other objects in strategic locations to create physical and visual stability.

banding. A design technique in which a group of stems is tied or bound together in one or more places using raffia, ribbon, or vines for decorative purposes.

banking pin. Straight silver pins, 1 to 2 inches long. Commonly used to pleat ribbon or pin flowers on hard foam. The term comes from the early use of such pins for wrapping currency.

Baroque period. A powerful and imaginative art and design direction during the 17th and 18th centuries in Europe characterized by elaborate and massive decorative elements and curved rather than straight lines. A reaction against the severe classic style. Floral arrangements typical of the Baroque period are tightly massed and overflowing, displaying a rhythmic asymmetrical balance.

basing. The application of materials over a design's base or foundation to camouflage floral foam and other mechanics. A variety of techniques may be used for basing arrangements, such as clustering and layering; this technique provides a foundation that adds color and texture to a composition.

basket bouquet. A specialty hand-held bouquet constructed in a basket.

bas-relief. Sculpture in which figures are carved in a flat surface so that they project slightly from the background.

beauty clip. Plastic U-shaped gripping device used to fasten chicken wire to the edge and around the perimeter of a container.

bent neck. A condition found on cut roses, gerbera, and other flowers when water no longer enters the stem, causing the stem beneath the flower to become weak, resulting in nodding flowers with limp stems that bend over.

Bible bouquet. A specialty bouquet carried by some brides; a Bible or prayer book accented with flowers and ribbon.

Biedermeier design. A design style influenced by a period in German and Austrian history. This style utilizes compact concentric circular or spiral patterns in a rounded bouquet.

binding. The process of tying similar materials together in bunches.

binding point. The point or area where all stems come together or intersect, as in a hand-tied bouquet.

biocide. Also called germicide. A general term for a chemical substance that can kill living organisms, especially harmful microorganisms that grow in vase water. An ingredient in floral preservative that inhibits the growth of microorganisms.

blade. The broad, flattened part of a leaf.

blockage. The clogging of water-carrying vessels in flower stems with air, sugars, salts, or bacteria that inhibits water uptake. Usually occurs at the base of the stem.

blueing. Purplish or bluish coloring that develops on flowers as a result of senescence or cold damage.

bosom bottles. Popular during the Victorian era, small containers made to hold flowers worn in the décolletage.

bostryx. A cymose inflorescence with successive branches on one side only, normally coiled like a spring.

botanical design. A design style that symbolizes the natural life processes or life cycle of a plant; often features bulb flowers.

botanical name. The two-part Latin name given to plants and flowers, consisting of the genus and species name. These names are used and known worldwide. Often called the Latin name, scientific name, or universal name. See *genus, Latin name,* and *species.*

botanicals. Everlasting flowers patterned after real flowers and foliage to be botanically correct and realistic in color, shape, texture, pattern, and size.

botrytis. *Botrytis cinerea* is a gray mold or fungus present in the air that may form on flowers causing irreversible damage. It is encouraged by high humidity and high temperature.

bough pot. A vase for branches or cut flowers. Often refers to the large containers used to hold flowers set in the fireplace during summer months.

boutonniere. A single flower or cluster of flowers worn on a man's lapel.

bract. A modified, usually reduced, leaflike structure that subtends a flower or inflorescence in its axil.

broad-leaf foliage. Large or wide leaf, such as the leaves of salal, ivy, and camellia.

broker. An individual who handles functions between growers and wholesalers, including the location of sourcing and coordination of delivery from all over the world.

bud cut. Flowers harvested in the bud stage, prior to the petals opening. Many flowers are harvested when the flower is tight in the bud form, such as roses and iris, allowing for a longer vase life.

bud-opening solution. A chemical solution containing sugar and a germicide to speed the opening of bud-cut flowers before they are sold; used in conjunction with warm temperature and high light intensity.

bud stage. Refers to the stage at which some flowers are harvested, at an immature stage when the blossom is tight and unopened; generally single or solitary flowers, they could continue to open to their full potential. Examples include roses, peonies, daffodils, iris, and tulips.

bud vase design. A floral design, usually consisting of one or several flowers, foliage, and fillers, made in a small vertical container vase.

buffering capacity. The water's ability to resist change with the addition of chemicals, particularly citric acid. Highly buffered water resists changes in pH while poorly buffered water is responsive to change.

bulb. A specialized underground plant organ. A general term for flowers that are produced from bulbs, such as tulips, daffodils, and hyacinths.

bunching. Packing flowers in groups of 10, 12, or 25 stems; or grouping by weight or the number of blossoms that prepares flowers for shipping.

bundling. The wrapping or tying together of materials.

Byzantine. Referring to *Byzantium,* an ancient city, and the *Byzantine Empire* in southeast Europe and southwest Asia (A.D. 395–1453). Byzantine floral designs reflect the decorative style of the mosaics. Symmetrical, stylized tree compositions were introduced during this time, as evidenced through mosaics.

cage holder. A stem-anchoring device with numerous openings in the surface, generally surrounding wettable floral foam, that provides a method for supporting cut flower stems.

calyx. Collective term for all the sepals of a flower.

candelabra. A fixture designed to support one or several candles for lighting; often decorated with flowers and ribbon bows.

canopy. An arch or covering used at weddings, often decorated with flowers and used at the altar, the rear of the center aisle, and the reception.

carbohydrate. An organic compound—includes sugars, starch, glycogen, and cellulose. Food that is produced by the plant during growth and serves the plant as a source of energy to keep it alive. Carbohydrates are the primary ingredients in floral preservatives.

care and handling. The procedures that involve attention and techniques directed toward increasing the longevity of fresh flowers and foliage. When proper care and handling is carried out at each level of distribution, vase life is increased for the final consumer.

care tags. Paper or plastic tags attached to floral arrangements and plants that provide appropriate information (name of flower or plant, care, etc.) for enhancing quality and increasing longevity.

carpel. One of the flower's female reproductive organs, comprised of an ovary and a stigma and containing one or more ovules.

cascade bouquet. A hand-held wedding bouquet style in which the flowers hang down (cascade) below the main portion of the design.

cash and carry. Merchandise paid for with cash and then taken by the customer. No or little added service, such as design and delivery, is given.

casket blanket. A blanket of flowers constructed on heavy fabric, such as burlap, and displayed by draping over the casket like a blanket.

casket inset. A small arrangement of flowers designed to be placed in the opened lid of a casket.

casket scarf. A design of flowers, often attached to fabric, that drapes over a part of the casket (from side-to-side or end-to-end).

casket spray. A floral arrangement that is placed on top of the casket during a funeral service.

catkin. A pendulous, spikelike inflorescence of simple, usually unisexual flowers; found only in woody plants; also called ament.

ceramics. Decorative or functional items and containers fashioned from clay and fired (baked in a kiln); includes earthenware or pottery, stoneware, and porcelain.

chain of life. A marketing and educational program sponsored by the Society of American Florists, specifically focused on proper care and handling at every level (link) in the channel of distribution (chain), ensuring maximum longevity and increased quality for the final consumer.

chaplet. A wreath or garland for the head, customarily made from flowers and foliage. Introduced in ancient times.

chemical solutions. See *conditioning.*

chenille stem. A tufted, velvety, wired stem, generally sold in 12-inch long pieces; available in a variety of colors and thicknesses; may be shaped and used as an accessory in designs.

chicken wire. A pliable fencing wire used in floral arrangement as an aid for holding flowers or helping heavy, thick-stemmed flowers stay in position while in the floral foam.

chilling injury. Plant disorder caused by low, but above-freezing, temperatures. Tropical flowers often experience chilling injury when exposed to temperatures below 47–50°F. Flowers fail to open properly, lesions appear on flowers and bracts, and stems and bracts may blacken.

Chinese style. A respected floral art form featuring unstructured, naturalistic designs using seasonal plant material. Arrangements are symbolic and generally include the use of a dominant vertical element and also the use of a more delicate horizontal element.

chlorophyll. The green pigment of plant cells necessary for photosynthesis.

chroma. A measure of the intensity or purity of a hue, also called saturation. The relative brilliance or dullness of a color.

chuppah. A canopy under which a Jewish couple is married.

circular design. Includes various rounded styles—both symmetrical as well as asymmetrical, and one-sided and all-sided; some are more mass designs, such as oval, fan, and topiary ball, while others are curvilinear such as crescent and Hogarth.

circular form. A spherelike appearance. All materials radiate out from a central area within the container.

citric acid. A natural chemical in citrus fruits. An ingredient of most floral preservatives, a compound used to lower the pH (acidity) of water for cut flowers. Used to create a hydrating solution for flowers.

classic Japanese design/style. An early Japanese style of design, focusing on nature and symbolism; also called *formal*, this style includes *rikka* and *shoka* or *seika*.

classic design/style. A timeless design, generally formal and balanced, restrained and simple; also called traditional, these designs are generally mass and full that continue in popularity and fashion; includes circular, oval, triangular, and fan-shaped designs; contemporary applications include *mille fleurs,* Biedermeier, phoenix, and waterfall designs.

clearinghouse. The head or central office of a floral wire service where transactions between florists are processed and accounts are balanced.

cleats. Crosswise strengtheners built into fiberboard boxes for shipping cut flowers.

clustering. A technique of gathering like materials together so closely that the quantity and shape cannot be determined; individual materials lose identity and function as one unit, emphasizing color and texture.

clutch bouquet. A simple or casual gathering of flowers tied with ribbon, raffia, or string for a natural, garden-picked appearance.

clutch wiring. A wiring method used to secure clusters of tiny flower stems and fillers; also called *wrap-around* wiring.

collarette. A thin cardboard support placed behind gardenia flowers to protect the fragile petals.

collaring. A method of pre-greening an arrangement, especially on round, circular designs by inserting foliage in the floral foam near the container rim all the way around, before inserting flowers; helps set up the circular shape.

colonial. A term loosely used in referring to the two-hundred-year period that includes the settlement of the early colonies in America through the Federal period (about 1607–1835); also called Colonial Williamsburg. Colonial style bouquets are typically rounded and massed, often combining fresh and dried flowers.

colonial bouquet. A hand-held bouquet style that is constructed in a circular shape.

color. The visual interpretation of light waves from the visible spectrum.

color wheel. The standard color wheel or circle has twelve colors, including primary, secondary, and tertiary colors; a simple and useful tool for the designer.

commentator. A person who reports, analyzes, and evaluates floral designs, floral industry products, and news at design shows and other industry events.

commercial account. A business account, such as with a hotel, computer company, or bank.

common name. The name by which a flower or foliage is generally known rather than by its botanical, scientific name.

complementary color scheme. The use of any two hues located opposite each other on the color wheel.

complete flower. In botany, a flower having four whorls of floral parts—sepals, petals, stamens, and carpels.

composition. A grouping or organization of different elements to achieve a unified whole.

compote. A raised, stemmed container.

compound. Consisting of several parts; a leaf with several leaflets, or an inflorescence with more than one group of flowers.

conditioning. The process of preparing flowers for shipping, storage, or arrangement. It usually involves treatment of flowers in a conditioning solution such as preservative water, silver thiosulfate solution, hydrating solution, or pulsing/sugar solution.

cone design. Typical to the Byzantine period style; a three-dimensional vertical isosceles triangle design that generally requires a foam base for mass flower insertion.

conifer. A cone-bearing tree, generally characterized by needlelike foliage.

consumption. Use and enjoyment of flowers, floral products, and floral services by the final consumer.

contemporary design/style. Those arrangements that are currently in fashion, popular, and representative of leading trends in creativity. A generic term for whatever is the current trend or on the leading edge in floral design.

continuation. Also referred to as transition, a method of achieving unity in design by planning a gradual change from one element to another, causing continuous eye movement.

contrasting color scheme. Based on unrelated colors from distant parts of the color wheel; visually exciting with great variety; examples include complementary, split complementary, and triadic.

cool colors. Blues, greens, and some purples, and colors containing these hues; associated with water and ice; generally restful, peaceful, and soothing. Cool colors are receding and fade into the background.

co-op. The association of several growers, drivers, and so on who offer merchandise or services under one roof or in a joint effort.

corner piece. A small arrangement of flowers designed to be displayed on the outside corner at the head of the casket.

cornucopia. A basket or other container shaped like a horn or cone overflowing with fruit and vegetables, flowers, foliage, and grain. Introduced during the Greek period, the cornucopia is known as the symbol for abundance. Also referred to as a *horn of plenty.* Floral arrangements are often made in cornucopia containers during the autumn, particularly for Thanksgiving.

corolla. All the petals of a flower; normally conspicuously colored.

corsage. A grouping of flowers, ribbon, and other accessories worn by a woman.

corymb. A rounded or flat-topped inflorescence of racemose type in which the lower (outer) flower stalks (pedicels) are longer than the upper (inner) ones so that all the flowers are at about the same level.

crescent bouquet. A hand-held cascading bouquet in the shape of a distinctive C-shape or quarter moon shape.

crescent design. An asymmetrical floral arrangement in the shape of the letter C; often used in pairs to accent something placed between them.

crescent form. The organization of floral materials in a C-shape or quarter moon shape.

C-shape design. See *crescent form.*

cultivar. A variety of plant found only under cultivation.

curved line. A swerving line; suggesting a natural, sweeping motion; adds interest and gentleness to floral designs.

custom design/service. Floral arrangements and service designed and provided specifically for the customer.

customer base. Individuals and commercial accounts who patronize or buy on a regular basis and not just for holidays.

customer bonding. Development of a close friendship between patronizers and the flower shop.

cyme. An inflorescence in which each terminal growing point produces flowers.

cytokinins. Class of plant growth substances that promote cell division, among other effects; added to some floral preservatives.

décolletage. (day-call-eh-TAZH) The neckline or top of a dress cut low to bare the neck and shoulders.

delftware. Glazed earthenware, also called delft, usually blue and white in color. It originated in Delft, a city in west Netherlands. A delftware brick is a rectangular box with a perforated top or grill for flowers. Originally produced in the late 1500s it again flourished into the mid-1700s. Delft of the early 1600s imitated Chinese porcelain.

delivery area. Space in a flower shop, usually at the rear of the store, where outgoing orders are prepared for delivery (extra packaging, care tags, and cards may be added, water levels, and addresses are checked); designs are loaded into the vehicle for delivery from this area of the shop.

delivery cooperative (co-op). A joint effort by several floral shops to deliver flowers to several areas.

depth. The distance from the front to the back of a design and the viewer's awareness of this distance. Depth may be emphasized through various techniques.

desiccant. A substance that absorbs moisture, such as sand, alum, cornmeal, and silica gel, used in drying flowers.

design. A planned organization of elements to suit a specific purpose.

diagonal line. A slanted line giving the feeling of mobility or instability, creating dynamic movement or tension.

dichasium. A simple cyme; an inflorescense having a central older flower and a pair of lateral branches bearing younger flowers.

directional facing. Turning or directing flower heads certain ways in order to increase interest and visual movement within a design.

direct mail. Advertising and promotional materials sent by mail to customers.

disc flower. Tubular flowers that compose the central part of a head flower in most *Asteraceae*, such as chrysanthemums, gerberas, and the like, contrasted with ray-shaped flowers on the margins of the head flower.

discordant color scheme. Using four colors that are widely separated on the color wheel. Care in color proportion and hue intensity and value must be taken when using unrelated hues. Examples of discordant schemes include double complement, alternate complement, and tetrad.

dish garden. A container with soil in which several different plants grow together; often decorated with accessories or cut flowers in water tubes inserted into the soil.

display. To visibly show merchandise to the public.

display cooler. A cooler in the sales area of a floral shop that displays flowers and arrangements for sale.

distribution channel. Route by which floral products move from grower to final consumer. Traditionally the channel of distribution involves the grower, auction, broker, wholesaler, retailer, and final consumer.

dixon pin. Two wooden picks attached on opposite ends of a flexible metal strip; used as a mechanical aid in floral work, particularly large sympathy designs.

domestic market. Products that are produced and consumed within the same country. For example, carnations grown in Colorado that are sold and shipped to buyers within the United States and consumed within the United States are a domestic product grown for a domestic market.

double complement. A discordant color scheme using two pairs of complementary colors.

double-end easel spray. A funeral design placed on an easel that is symmetrical and radiating and has flowers not only in the upper portion of the design, but in the lower portion as well.

dry-pack. Storage or shipping of flowers out of solution (in a dry manner). Proper humidity and temperature conditions are essential for the success of dry-pack storage and shipping.

dry prone. Flowers and foliage that dry out easily and are subject to wilting, shedding of leaves, needles, and florets. These flowers and foliage must be taken care of first and conditioned quickly on arrival at the wholesaler or retailer.

Dutch-Flemish style. A style of floral arranging copied from the paintings of the Dutch and Flemish artists of the 17th and 18th centuries. Typical floral arrangements are massed and overflowing with the use of many varieties and colors of flowers (including tulips) facing in all directions. For authenticity in recreating this style of design, many accessories are generally placed around a lavish floral bouquet, including fruits, bird nests, and shells.

dynamic line. An active line of continuous movement that counters a design's shape, such as a diagonal line contrasted with horizontal and vertical lines within a rectangular format.

Early American style. Floral arrangements made in the Early American style are simple and charming, using native plant materials, such as wildflowers, weeds, and grains. Representative of the Early American Period (1620-1720), containers are generally simple utility jugs, pitchers, and other kitchenware made from pottery, copper, and pewter.

earthenware. Containers, tableware, and other items made of coarse, brown or red clay. After firing, earthenware is porous and nondurable unless treated with a glazed finish. Generally, earthenware is coarse-textured and heavy.

easel. A three-legged stand constructed of heavy wire or wood used to support and display wreaths, other set pieces, and floral sprays, generally at funeral services.

easel spray. A floral arrangement placed on an easel.

eclecticism. In floral arranging, refers to borrowing and mixing styles from various sources and periods and combining them into one style of design with an eye to compatibility.

Egyptian period. Approximately 2800–28 B.C., the floral style was simplistic, repetitious, and highly stylized. Flowers and fruits were placed in carefully alternating patterns. Chaplets, wreaths, garlands, and flower collars were also popular.

emblem tribute. Also called organizational tribute, a funeral design or set piece ordered by members of a club, business, religion, or school to which the deceased belonged.

emphasis. A design principle synonymous with focal point. The creation of visual importance or accent in a design; the center of interest.

empty chair design. An emotionally expressive funeral set piece, also called the "vacant chair" made entirely out of flowers attached to a wire frame or to a real chair; this generally life-sized designed gained popularity in the early 1900s; it is still used today in isolated areas; the empty chair represents a missing individual and undying hope, as well as the loss of an authoritative person.

English-Georgian period. Also called *Georgian*, refers to the period in England during the reigns of George I, II, and III (1714–1790). Floral styles during this period included simple hand-held bouquets that were carried for fragrance. Floral arrangements were symmetrical and ranged from small to large with great varieties of fragrant flowers. A variety of containers was used, including Wedgwood, metal, and glass.

epaulet corsage. See *over-the-shoulder corsage.*

epergne (eh-PURN). An ornamental stand with several separate dishes or trays used as a table centerpiece for holding fruit and flowers; popular during the Victorian era.

ephemeral. Flowers lasting a very short time, such as cut gardenia blossoms.

equilateral triangle. A design form having the shape of a triangle with all three sides of equal length.

EthylBloc. The trade name for 1-methylcyclopropene (1-MCP) which is a gas in its natural form. EthylBloc is a powder that, when mixed with water, releases a nontoxic gas that blocks the detrimental effects of ethylene. Because 1-MCP is a gas, treatment needs to be carried out inside a sealed space (such as in a sealed truck or storage cooler). See *ethylene.*

ethylene. A natural plant hormone (C_2H_4) that is responsible for a range of problems, including abscission of flowers and leaves, leaf yellowing, and flower distortion. It is a colorless, odorless gas that hastens senescence of flowers. Called the "aging hormone," it is emitted by ripening fruit, aging flowers and foliage, equipment used around flowers, such as heaters and forklifts (exhaust from internal combustion engines), and cigarette smoke.

European style. A loose term generally referring to full, massed bouquets that use a great variety of flowers and colors, in contrast to the *oriental style.*

everlasting flowers. Dried or artificial flowers as contrasted with fresh flowers.

ewer. A large water pitcher with a wide mouth, often used for holding flowers.

exotics. A general term referring to tropical, unusual, and form flowers and foliage.

experimental design. A general term for new, contemporary styles of arrangement, often bizarre in nature.

exporter. One who carries or sends cut flowers, foliage, and floral products to another country or other countries to sell.

faience. Glazed earthenware named for Faenze, Italy, where it was originally manufactured.

fan bouquet. A hand-held grouping of flowers, placed on a lace, decorative fan.

fan design. See *radiating design.*

fan-shape. A radiating, half-circle design.

feathering. The separation of a flower, especially a carnation, into several small florets for use in a corsage or bridal bouquet.

Federal period. The political, social, and decorative formation era in America following the Revolutionary War (1790–1825). Floral arrangements during this period were styled after ancient classic designs as well as elaborate European massed, symmetrical bouquets.

filament. The anther-bearing stalk of a stamen.

filler. A type of flower, foliage, or accessory used to fill in spaces in an arrangement, add interest, and complete a design.

filler flower. Generally has a complex branched system of stems and tiny florets; used to fill in empty spaces, add accent, and complete a design; examples include baby's breath, statice, aster, and waxflower.

filler foliage. Greenery used to fill in and complete a design; examples include plumosa, tree fern, sprengeri, and ivy.

filling florist. A florist who receives a flowers-by-wire order and fills it by designing and delivering the flowers to the specified recipient.

final consumer. The last person to receive flowers through the channels of distribution.

flags. Loops of ribbon with tails.

flat spray. A showy gathering of flowers and foliage, tied with a ribbon and used to decorate graves. In regions where flat sprays (also called bunches) are popular, many are grouped together to form a larger wall spray.

floral art. A creative form of expression using flowers and, but not always, a container, foliage, and accessories, while incorporating the principles of harmony and unity; a floral artist is one who displays expertise or great proficiency, and unusual perception in creating beautiful, stunning, extraordinary, and often bizarre floral designs.

floral clay. A waterproof, sticky material used in floral arranging for fastening stem-anchoring devices, such as needlepoint holders and hard foam, to containers.

floral designer. A person who designs or makes original floral arrangements; a talented, creative, and skilled florist.

floral foam. A highly porous material used to absorb water, support stems, and hold flowers, foliage, and accessories in place in a design.

floral holiday. A day that is associated with the giving and receiving of flowers, such as Valentine's Day and Mother's Day, or a day that is associated with using flowers for decorating, such as Christmas and Memorial Day.

floral industry. The particular branch of the business that includes the production, manufacturing, marketing, and selling of flowers and floral products.

floral pomander. A sphere or ball of fragrant flowers that hangs from ribbon.

floral preservative. A chemical mixture of sugar, biocide, and citric acid, added to water to extend the life of cut flowers.

floral rosary. A floral replica of the prayer beads used by Catholics; rosebuds are inserted into a special rosary string available from most floral supply houses; the floral rosary is generally a funeral inset design and is draped inside the open casket lid.

floral tape. A paraffin-coated paper that may be stretched over stems to securely bind them together when creating corsages, boutonnieres, hand-held bouquets, and everlasting designs.

floret. One of the small flowers that make up the total inflorescence.

floriculture. The cultivation and production of floral crops, especially those that are harvested and sold.

Florist Information Committee (FIC). A branch of the Society of American Florists that functions to alleviate any negative press about the floral industry and its products and instead provides positive information to the public.

flower. The reproductive structure of angiosperms; see *complete flower.*

flower merchandiser. A retailer specializing in loose cut flowers sold by the stem or bunch who generally does not provide design or delivery services.

flowers-by-wire. See *wire order.*

foam forms. Hard or wettable foam forms precut by the manufacturers in popular shapes, such as crosses, hearts, open hearts, and wreaths.

focal area. A part of a design that is emphasized by contrasted materials, the area of focus; generally in the base of a design where all materials meet.

focal point. The center of attraction, usually within the focal area; often highlighted with a distinctive flower or accessory item.

football mum corsage. A traditional corsage worn at the high school or college homecoming football game, dance, and other associated events; generally designed with one or several standard incurve chrysanthemums and a variety of accessories, such as tiny footballs and megaphones, long ribbon streamers in the school colors, and the school's initials or insignia.

foreign market. Products that are produced in one country and sold to another for consumption. For example, carnations grown in Colombia that are sold and shipped to buyers within the United States and consumed within the United States are a foreign product grown for a foreign market.

form. A design term synonymous with shape or outline.

formal. A general term for the classical, symmetrical, or elaborate.

formal linear design. A high-style design emphasizing shape, form, and line, generally with asymmetrical balance. Minimal materials are used, emphasizing individual flowers, stem angles, colors, and textures.

form flower. Denotes an unusual or distinctively shaped flower; attracts attention and provides emphasis; examples include bird of paradise, anthurium, and orchids. See *exotics.*

form foliage. Greenery with an interesting shape that draws attention; often the leaves will possess unusual textures, colors, or patterns as well; generally used with form flowers; examples include caladium, croton leaves, monstera, papyrus, and cyperus.

foundation. The base of a design from which elements rise.

foundation colors. The primary colors—red, yellow, and blue—from which all other hues on the color wheel are created.

fragrance. The quality of being fragrant or sweet-smelling; an often overlooked element of floral design. When scented flowers are used in a bouquet, their fragrance adds another pleasurable dimension to our enjoyment.

framework. The basic or beginning structure of a floral arrangement, established with stems of flowers and foliage. The framework creates the pattern of a design.

framing. A design technique in which the perimeter of the design fully or partially encloses an area and focuses attention to its contents.

free-form design. A design that is not confined to any geometric shape and generally emphasizes line and texture.

freelance floral designer. A knowledgeable and experienced person who designs floral arrangements for individual buyers.

freeze-drying. A method of drying plant materials. Moisture is removed from the cell structure of flowers by mechanical means; flowers retain shape, suppleness, texture, and usually color.

French period. Also known as the *Grand Era,* 17th and 18th centuries in France during the reign of Louis XIV. This style was influenced by the Dutch Flemish style, with emphasis on classic form, refinement, and elegance, rather than overdone flamboyance. Highly ornamental vases were used to hold tall fan-shaped, rounded, and triangular bouquets. Often termed *French Rococo,* this style was extravagant, using symmetry, shells, rocks, and all manner of elaborate decoration.

frond. The leaf of a fern; also, large, divided leaf.

full-couch. A casket with an undivided full lid (contrasted with a half-couch).

full-service flower shop. A floral shop that provides every floral product and service needed by customers, including delivery and wire service.

fully open bud stage. Refers to the state at which some flowers are harvested, at a mature stage before beginning to shed pollen. If cut too early, these flowers do not continue to open to their full potential. Examples include zinnia, sunflower, marigold, and calendula.

funeral basket. An arrangement made in a container of wicker, plastic, papier-mâché, or metal, usually with a handle. It is displayed on the floor or a stand at funeral services.

garland. A wreath, woven chain, or festoon of flowers, leaves, or other materials worn on the head or used as decoration.

gates ajar design. An emotionally expressive funeral set piece that gained popularity in the early 1900s; the arch and gates, left slightly open, represent a recent departure of the spirit into heaven; this set piece remained popular for many decades resulting in a variety of adaptations, such as "the Golden Way" gates ajar design which had a pathway in the foreground leading up to the gates of heaven.

gauge. Standard scale of measurement used to determine the thickness of wire; the lower the number, the thicker the wire; common floral wire gauges range from heavy #16 to delicate #30.

genus, *pl.* **genera.** The taxonomic group between family and species that includes one or more species that have certain characteristics in common. See *botanical name, Latin name,* and *species.*

geometric design. A floral arrangement characterized by straight lines, triangles, circles, or similar regular forms.

Georgian period. See *English-Georgian period.*

geotropism. Bending and curving upward against the force of gravity, exhibited by gladiolus and snapdragons.

germicide. Any antiseptic used to destroy germs; see *biocide.*

glamellia. A wired and taped flower constructed from gladiolus petals and formed to resemble a camellia flower; used in corsages and bridal bouquets.

glamour leaves. Accessories of artificial foliage, made from fabric, net, lace, and so on, used in corsages.

glycerin. A solution used for preserving foliage and some flowers. When properly absorbed in plant stems, it keeps the plant material soft and pliable.

going-away corsage. A floral remembrance worn by a bride as she leaves with her husband for the honeymoon; sometimes a removable part of the bridal bouquet.

golden anniversary. The 50th anniversary. Floral arrangements for this special occasion are often made in gold containers, incorporate gold and yellow flowers, and often have a gold number 50 insignia accessory.

golden mean. A Greek rule of proportion referring to the division of a line somewhere between one-half and one-third its length that is the most pleasing to the eye.

golden rectangle. A Greek standard for proportion; a rectangle or oblong with its sides in a ratio of 2:3.

golden section. A Greek rule of proportion based on the golden rectangle. It involves the division of a line or form in such a way that the ratio of the smaller portion to the larger is the same as that of the larger portion to the whole. For example, 2:3 is about the same ratio as 3:5, 5:8, 8:13, and so forth. This rule may be applied to the ratio of the container to the flowers, and of the flowers to the entire arrangement.

gradation. A design technique of placing flowers or foliage in a sequence, from largest to smallest, darkest to lightest, and so on.

grading. Evaluating crop quality utilizing a standardized system of measurements of crop quality. For example, roses are graded according to the length of stem and overall quality.

Grand Era. See *French period.*

Greek period. Floral styles of this time (600–146 B.C.) were garlands and wreaths. Flowers were scattered on the ground during festivals. Fragrance and symbolism was of utmost importance.

Greek revival. The final phase (1825–1845) of the neoclassic style in America; floral designs were generally large and symmetrical; containers reflected the classic styles of the Greek and Roman periods.

greening pin. U-shaped pins similar to hairpins; commonly used to secure moss to foam. Also called philly pins and fern pins.

green tape. See *floral tape.*

grid. Framework of plastic, tape, or wire used at the top of a vase to keep flowers in position without the use of floral foam.

groomsmen. The men who attend the groom at the wedding.

grouping. A design technique characterized by placing like materials in sections or groups with negative space between various groups and some space left between flowers within groups.

grower. The initial level of distribution of cut flowers and foliage and potted plants. Foreign growers sell floral products directly to brokers and wholesalers or take their products to an auction, while domestic growers most often sell to wholesalers. Proper care and handling, or the chain of life, begins at the grower level.

growth regulator treatment. The special hormone-containing solutions frequently used by growers to prevent and inhibit problems, such as leaf yellowing on alstroemerias and lilies.

gynoecium. The collective term for the female parts of a flower, the pistil or pistils.

half-couch. A casket in which the lid is divided in half. The head of the casket is open for viewing, while the foot of the casket is closed.

half open bud stage. Refers to the stage at which some flowers are harvested, when some of the florets are open and some are still in the bud stage; generally refers to spikelike flowers and other inflorescence types. Examples include gladiolus, delphinium, snapdragons, agapanthus, and lilies.

hamper. A tall container made of wooden slats or corrugated cardboard used to ship tall flowers, especially those that are geotropic, such as gladiolus.

hand-tied bouquet. A bouquet that is made in one hand while the other hand places flowers and foliage in a spiral pattern within the bouquet. It is generally tied with ribbon, string, or raffia.

hardening. The conditioning process that allows flowers to become hard or turgid from water uptake.

hardgoods. Nonperishable staples or inventory items.

harmony. A pleasing relationship among the parts and elements of a design.

harvesting. The cutting of flowers and foliage.

head flower. A dense inflorescence of small, crowded, often stalkless flowers; a capitulum.

head table. A separate table (often at the head or front of the room) where the bridal party sits at a lunch or dinner; this table is the central element in the room; flower arrangements used on this table are generally more elaborate than those for the guest tables, however, they must be kept low, so the bridal party may be seen.

heika (HAY-kah). Vase flowers. General term for naturalistic Japanese arrangements in tall vases.

herbaceous stem. A general term referring to any non-woody plant. Also referred to as soft stem.

high style. Characterized by bold forms and lines. See also *art deco* and *formal linear design.*

Hogarth curve. A design shape or prominent line in a design in the shape of the letter S. Named after William Hogarth who theorized that all beauty was based on a serpentine S-line.

Hogarth design. A curvlinear floral arrangement in the shape of the letter S; the serpentine line displays beauty and grace. See *Hogarth curve.*

holding solution. Preservative solution used in storing cut flowers.

hook-wiring. A method of wiring flat-topped flowers, such as chrysanthemums using a U-shaped wire with a hook inserted down into the center of a flower and into the stem.

horizontal design. A floral arrangement that has a strong long and low line emphasis, parallel with the tabletop; often used as a centerpiece and can be symmetrical or asymmetrical and one-sided, or all-sided.

horizontal line. A flat line, parallel to the tabletop or floor. Horizontal lines provide stability and restfulness.

horn of plenty. See *cornucopia.*

horticulture. The science and art of producing and using ornamental plants, fruit, and vegetables.

HQC. 8-hydroxyquinoline citrate. An anti-microbial agent or biocide that kills bacteria and other organisms.

hue. The property that gives a color its name, such as red, blue, yellow, or green.

hyacinth stake. A slender, wooden stick used to provide additional support to flowers.

hybrid. The offspring of two parents that differ in one or more heritable characteristics. The offspring of two different varieties or of two different species.

humidity. Refers to the amount or degree of moisture in the air. See also *relative humidity.*

hydrating solution. A citric acid solution that encourages rapid uptake of water and helps flowers recover from dry shipping, dry storage, and water stress.

ikebana (ee-keh-BAH-nah). Literal translation: living flowers. General term applied to all Japanese flower arrangements.

Ikenobo (ee-kee-NO-bah). Literal translation: hut by the pond. Supposedly referring to the priests' place of retreat in the temple grounds, it is the original term applied to flower arrangements, later used to denote a school. The Ikenobo school, the oldest school of flower arranging in Japan, had its beginnings in the 6th century.

implied line. An imaginary line created by a series of repetitious elements.

importer. One who brings in cut flowers, and foliage, and floral products from another country or countries for purposes of sale.

inflorescence. Any arrangement of more than one flower on a stem, for example, corymb, cyme, and panicle.

informal. Casual and relaxed; often asymmetrical and natural.

in lieu of flowers. An addition in death notices asking people not to send flowers to a funeral.

insertion wiring. Method of wiring fresh flowers whereby wire is inserted up through the base of the stem all the way to the top to provide support; often used for gerberas and other hollow- or fleshy-stemmed flowers.

intensity. See *chroma.*

intermediate colors. See *tertiary colors.*

interpretive design. A floral composition that expresses the designer's feelings and ideas.

inventory. Merchandise and stock on hand. A listing of the stock within the store.

isolation. Separation of a flower or group of flowers from the rest of the arrangement. A method of creating an area of emphasis or a focal point.

isosceles triangle. A triangle form that has two equal sides and one side different in length. Usually a tall, symmetrical triangular floral arrangement.

Japanese design/style. General term for a linear floral design characterized by three main lines or sections.

jiyu-bana (jee-yoo-BAH-nah). A freestyle Japanese floral arrangement.

juxtapose. To place side-by-side or close together.

kenzan (KEHN-zahn). A needlepoint holder used in low bowls.

kubari (koo-BAH-ree). A forked twig used as a flower fastener in a taller container used for classical Japanese designs.

lacing. A natural method of crossing stems to form a grid to hold flowers in position within a vase. A technique for making hand-tied bouquets.

landscape design. A floral arrangement that depicts a large area of nature. Flowers, branches, and foliage may represent parts of a natural landscape or groomed garden.

latex. A milky and usually whitish fluid that is produced in the cells of various plants. Bleeding latex of cut flower stems must be cared for properly.

Latin name. The scientific or univeral name of a plant; although plants are generally derived from Latin words, they may also be derived from Greek or personal names, as well as from other languages (whatever their origin, all names are treated as Latin). Plants were first given scientific names in the 16th and 17th centuries when Latin was a common language. Today, the Latin name is a universal name and forms a method of communication between nations and at all levels of the floral industry. See *botanical name, genus,* and *species.*

layering. The process of compactly overlapping like materials (usually leaves, such as galax) so that no or little space is left between them. This process produces a scale-like appearance.

leaflet. One of the parts of a compound leaf.

lei. A wreath, garland, or ornamental headdress or necklace made of leaves and flowers.

le style 25. See *Art Deco.*

liknon. An ancient basket in Greek and Roman times used for holding flowers.

limited-service flower shop. A flower shop characterized by little or no added service and products; for example, a floral department in a grocery store.

line. A continuous pathway for the eye to follow.

line/linear design. A general term for a floral arrangement characterized by strong vertical, horizontal, diagonal, or curving lines with negative space between for emphasis; contrasts with mass design; contemporary or advanced styles include western line, parallel systems, new convention, and formal linear designs.

line flower. A spike or spikelike inflorescence with an elongated stem, such as gladiolus, liatris, larkspur, and foxtail lily; adds height, width, and depth; easily sets the framework, shape, and size of an arrangement.

line foliage. Tall and long foliage having a linear shape; often used in conjunction with line flowers to set the framework of a design; adds height and width; examples include Scotch broom, gorse, myrtle, New Zealand flax, and pandanus.

L-shape design. See *right triangle.*

management. The act, art, or manner of managing, or handling, controlling, directing employees, designs, inventory, money, and various business affairs.

marketing. All business activity involved in the moving of goods from the grower or producer to the final consumer, including selling, advertising, packaging, and so on.

markup. A system for determining the sale price of an item based on the cost of merchandise.

mass design. A general term for a full, overabundant floral arrangement. This design contrasts with line design.

mass flower. A rounded flower at the top of a stem. Used for adding mass or fullness to a design; adds bulk and weight; examples include carnations, roses, chrysanthemums, and tulips.

mass foliage. Adds weight, bulk, and fullness to a design; efficient in covering up the floral foam and other mechanics of a design; examples include leatherleaf, salal, and pittosporum.

mass merchandiser. One who promotes, advertises, and organizes the sale of flowers in large quantities, such as grocery stores for their floral departments.

mechanical aid. Any material used in a container or in the designing of an arrangement to ease and expedite construction.

mechanical balance. See *balance.*

mechanics. The method of construction of a floral design; the technical aspect.

merchandise. Material offered for sale.

metabolism. The sum of all chemical processes occurring within a living cell or organism.

Middle Ages. The period of European history between ancient and modern times (A.D. 476–1450). Also known as the Medieval period and Dark Ages. Little is known of floral art during this time. However, fragrant flowers were highly favored for strewing on the ground, freshening the air, and for making wreaths and garlands.

mille fleurs design. Literal translation: thousand flowers. General term for a mass floral bouquet with many colors and flowers juxtaposed in an arrangement.

mirroring. Repetition of the like materials at varying heights and depths. See *shadowing.*

modernistic design. A contemporary floral arrangement expressing current trends and ideas.

monochasium. A type of inflorescence in which there is a single terminal flower, and below this a single branch bearing one or more younger flowers; variations of coiled and zigzag patterns inlcude helicoid, cincinnus, bostryx, and scorpioid.

monochromatic color scheme. Uses one hue from the color wheel and may include its tints, tones, and shades.

moribana (moh-ree-BAH-nah). Literal translation: piled-up flowers. General term applied to naturalistic Japanese arrangements in shallow or low, flat containers.

muff bouquet. A specialty bouquet in which a small floral arrangement is pinned to a muff for winter weddings.

nageire (nah-GAY-reh). Literal translation: thrown in. A loose, naturalistic style of Japanese arrangement of Ikenobo. General term for arrangements in tall vases or baskets.

naturalistic design. A design style that emphasizes the beauty of flowers and plant materials without manipulation.

naturalistic Japanese design/style. A style of Japanese design that is more natural and casual when compared with other Japanese styles; also called *informal,* this style includes *nageire* and *moribana.*

needlepoint holder. See *pin holder.*

negative space. Empty areas between flowers or materials.

neoclassic style. Revival simulating the ancient classical designs of Greece and Rome; influence by the French style of design; floral designs are generally symmetical, tall, and large, often pyramidal or fan-shaped; noted for using formal, elegant, and ornate containers. See also *Federal period* and *Greek revival.*

netting. See *tulle.*

networking. The exchanging of information with others in an informal way to further business.

neutral colors. Colors that are not located on the color wheel but that influence those that are. Generally black, white, and gray. Other neutral colors include tan and ivory.

new convention design. A design style characterized by only vertical and horizontal placements of materials, all of which are at a 90-degree angle to one another.

new wave design. A style of design in which materials are used in a nonrealistic way. Materials are painted, folded, and manipulated, while lines cross and zigzag throughout the design.

node. The part of a stem, such as on a carnation, where one or more leaves are attached.

nomenclature. The system of naming plants and flowers.

nosegay. A tight grouping of flowers, herbs, and foliage in a hand-held bouquet. Also called a tuzzy-muzzy, or tussie-mussie, and posie bouquet.

novelties. Unusual articles that are chiefly decorative and often used with flowers to suggest a floral holiday, season, or theme.

novelty design. A unique, one-of-a-kind floral arrangement designed for a certain situation or a special occasion; made with short-stemmed or broken flowers; popular examples include ice cream sodas, cakes, kittens, dogs, and caterpillars; accessories can be glued onto flowers to add detail.

object line/stem. Designation applied to earth line in the Ohara school.

Ohara. Twentieth century school of Japanese flower arrangements. Originator of the *moribana* style; interprets the *moribana* and *nageire* styles.

olfactory. Relating to the sense of smell.

one-sided design. A floral arrangement designed to be viewed from one side only.

open balance. A type of balance often employed in some contemporary design styles. A relaxed and unstructured balance; a feeling of stability resulting from no formal rules of construction.

open house. Promotion to get the general public to visit the floral shop to see the displays, usually prior to a major floral holiday.

open order. An order for a floral arrangement in which the designer decides both the contents, style, and price.

orchid grass. Shredded waxed paper used in packaging corsages, boutonnieres, and other small designs.

order-taker. A passive salesperson who simply writes up the orders for the customers without any salesmanship involved.

organizational tribute. A special easel set piece patterned after an organization's insignia or emblem; a funeral tribute given by members of the club, school, business, or religion to which the deceased belonged.

oriental style. A loose term referring to line designs including both the Chinese and Japanese styles, as well as designs that resemble these styles. The oriental style uses few materials and emphasizes simplicity, form, line, and texture.

ovary. An enlarged basal portion of a carpel or of a gynoecium composed of fused carpels. The ovary becomes the fruit.

overhead. Expenses and the general cost of running a business.

overlapping. A method of enhancing depth by varying the placement and positioning of flowers and foliage; placing one element on top of or in front of another.

over-the-shoulder corsage. An elegant corsage made to be worn on top of the shoulder and cascade down both front and back; also called an epaulet corsage.

ovule. The structure that becomes a seed after fertilization.

packaging. Wrapping or boxing flowers, plants, and floral products for sale and delivery.

packing. The large-scale, commercial processing and packaging of flowers and floral products in boxes for shipping.

paddle wire. Continuous wire wrapped around a small length of wood.

pall. A fabric piece placed on a casket. A floral design placed on a casket.

panicle. A branched raceme with each branch bearing a further raceme of flowers. More loosely applies to any complex, branched inflorescence.

pan-melt glue. Small pillows or pellets of glue that are melted down in an electric pan and used to secure mechanical aids and other items in designs.

parallelism. Placing two or more stems, lines, or other items in the same direction (usually in a vertical format) where the lines never meet.

parallel systems design. A style of design in which two or more (usually vertical) groupings are placed in a container and negative space surrounds the groups. Each group generally uses like materials.

parasol bouquet. A specialty bouquet in which flower clusters adorn an open or closed parasol (decorative, lacy umbrella).

pavé (pah-VAY) design/technique. A style of design or a technique in which materials are placed closely together in a cobblestone effect. The term is borrowed from the jewelry-making technique in which gems are placed closely together with little or no metal showing between them.

pedicel. The stalk of a single flower in an inflorescence.

peduncle. The stalk of an inflorescence or of a solitary flower.

perianth. The petals and sepals combined.

period style. A term used to designate a single item or a complete arrangement style prevalent in a specific country at a particular time in history.

perishable. Subject to decay; short-lived.

perishables. Generally refers to the fresh cut flowers, foliage, berry branches, and so forth, that are subject to deterioration and spoilage; in contrast to *hardgoods*.

petal. A flower part, usually conspicuously colored. One of the units of the corolla.

petiole. The stalk of a leaf.

pH level. A symbol denoting the relative concentration of hydrogen ions in a solution. pH values run from 0 to 14, and the lower the value, the more acidic a solution. For example, pH 7 is neutral, less than 7 is acidic, and more than 7 is alkaline.

phloem. Food-conducting tissue.

phoenix design. A style of design in which tall materials burst out from the center of a rounded design. Named for the ancient mythological Egyptian bird.

photosynthesis. The conversion of light energy to chemical energy. The production of carbohydrates from carbon dioxide in the presence of chlorophyll by using light energy.

phototropism. Turning or bending in response to light. Many cut flowers curve toward a light source.

physical balance. See *balance.*

physiology. The study of the activities and processes of living organisms.

phytohormone. Natural hormones or growth regulators produced by living plants that help delay or speed the aging process.

pierce wiring. A method of wiring flowers using a straight wire; the usual method for adding a wire to roses and carnations.

pigment. In plants, the substance that absorbs light, expressing colors.

pillow design. An emotionally expressive set piece that gained popularity in the early 1900s and in some areas is still commonly displayed at viewings and funerals; it is symbolic of peace, comfort, and eternal rest, as well as family, love, and home; flowers, ribbon, and fabric may be attached to a pillow form or frame in a variety of ways.

pillowing. A clustering technique used at the base of a design in which pillows or rolling hills and valleys are formed by the groups of flowers.

pin holder. A holder made of steel needles used to hold flowers and other plant materials in place in a low container.

pistil. Central organ of flowers typically consisting of ovary, style, and stigma. A pistil may consist of one or more fused carpels.

plugging. A general term referring to impeded uptake of stems caused by bacteria, salts, air, and the like.

pollen. A collective term for pollen grains.

poly foil. Colored aluminum foil used to wrap potted plants.

pomander bouquet. A specialty bouquet that is spherical or ball-shaped, often highly fragrant, and hangs from a ribbon, cord, or rope; sometimes made for brides, bridesmaids, and flower girls; may be used as a hanging decoration; sometimes called a *flower ball* or *kissing ball.*

positive space. Area occupied by flowers, foliage, and other materials.

postharvest physiology. The study of the activities and processes of harvested or cut flowers and foliage.

postharvest technology. The science or study of the plant processes, practical and helpful methods, and explanations regarding the care and handling of flowers and foliage during the period following harvest; new information regarding cut flowers and foliage dealing with extending the length and quality of vase life.

posy holder. A hand-bouquet holder made from various materials. Manufactured to hold tied nosegays, making them less cumbersome and longer lasting.

prayer book bouquet. See *Bible bouquet.*

precooling. The practice of cooling flowers after boxing by forcing refrigerated air through holes in the walls of the box.

pregreen. To fill containers with wet floral foam and greenery before flowers are added. Containers are often pregreened in mass quantities prior to floral holidays.

presentation bouquet. See *arm bouquet.*

preservative solution. A mixture of floral preservative with water for increasing the life of cut flowers.

preserved flowers. Real flowers that have been dried and preserved by one of a number of drying methods.

pressing. A method of drying flowers and foliage resulting in flat forms.

pretreatment. Special treatments used prior to using floral preservative in the care and handling process, for example, silver thiosulfate treatment for ethylene-sensitive flowers; also known as supplements, special treatments, specific-action chemicals, and conditioners.

primary colors. Red, yellow, and blue. These three colors are equidistant on the color wheel.

processing. The methods of care and handling, specifically, recutting and conditioning flowers and foliage.

processional. The ceremonial preceding of the wedding party up the aisle prior to the wedding ceremony.

producer. One who produces or grows cut flowers and foliage for sale.

Professional Floral Commentators International (PFCI). A division of the Society of American Florists established to promote communication and understanding within all segments of the floral industry.

promotion. All advertising, publicity, and personal selling activities leading to public recognition.

proportion. A principle of design; the comparative size relationship between the parts of a floral arrangement to each other and the parts to the whole; flowers, foliage, container, and the surrounding space all relate to proportion.

prospects. Potential customers or flower buyers.

proximity. Near in space.

psychic line. A line that is felt between two elements, created by placing flowers and materials in a way that directs the eye.

pubescent. Covered in soft, short hairs. Coarse in texture adding interest to designs.

publicity. Free promotional activities resulting in public notice.

public relations. The act of promoting goodwill between a business or person and the public.

pulsing. A postharvest technique used to load flowers with sugar and other chemicals prior to shipment.

queen's bouquet. See *arm bouquet.*

raceme. An inflorescence consisting of a main axis bearing single flowers alternately or spirally on stalks (pedicels) of approximately equal length.

radial balance. A type of balance or feeling of stability created by all elements radiated or circling out from a common central point like the spokes of a wheel or the rays of the sun.

radiating design. Also called fan-shaped arrangement; these designs are one-sided and similar in shape to a half circle with radiating lines converging from a central location.

radiating line. Lines or stems that converge from a central location to form a focal point; a method of achieving increased visual eye flow or rhythm.

raffia. Fiber from the palm tree *Raphia ruffia,* used as string.

ratio markup. A method of calculating the selling price based on the wholesale cost of items; simply by doubling, tripling, quadrupling, and so forth, the wholesale price, by which the retail selling price may easily be determined; often a guessing game for florists whereby many items such as labor, profit, and operational expenses are not covered.

ray flower. Flattened, colored ray-shaped flowers on the margins of most *Asteraceae* flowers, such as in sunflowers, gerbera, and chrysanthemums; often incorrectly called petals.

receiving area. Space in the rear work area of a shop where flowers and other merchandise arrive and where flowers are processed.

receiving line. The line in which the bridal party members and parents stand to receive guests' congratulations and greetings after the ceremony or at the reception.

receptacle. That part of the axis of a flower stalk that bears the floral organs.

recessional. The ceremonial proceeding of the wedding party down the aisle after the wedding ceremony.

recut. The process of removing the lower one-half to one inch of stem with a sharp blade, shears, or underwater cutter.

regional holiday. A holiday that is celebrated and associated with the use (decorating, giving, or receiving) of flowers in one area, while in another area the same holiday is not associated with the use of flowers; examples include Memorial Day and St. Patrick's Day. See *floral holiday.*

rehydration. A process in which cut flowers and foliage (especially those that have been dry-packed and recently shipped) are immediately recut and placed in a chemical solution containing deionized water, citric acid, a germicide, and wetting agents so they will hydrate quickly.

related color scheme. Based on a common hue that acts as a unifying element; the colors forming a related scheme may be variations of only one hue or may be variations of several adjacent hues on the color wheel; visually harmonious but may be boring; examples include monocrhomatic and analogous.

relative humidity. The amount of water vapor actually present in the air at a given temperature as compared to the maximum amount the air could hold at that certain temperature. Warm air has the capacity to hold more moisture than cold air.

Renaissance. A period in Europe after the Middle Ages. Beginning in Italy in the 14th century, it was marked by a humanistic and classical period in which an unprecedented flourishing of the arts occurred. Floral arrangements characteristic of the Renaissance period were massed in tight symmetrical shapes. Colorful flowers were often combined with fruits and vegetables. Many types of containers were used including urns, jugs, and bowls. Also well known was a single stem of white lily (Madonna lily) in a simple container. Flowers were symbolic with religious themes.

repetition. A method of obtaining rhythm by repeating similar elements throughout a design.

respiration. An intracellular process in which food is oxidized with the release of energy, the complete breakdown of sugar or other organic compounds to carbon dioxide and water. Respiration continues in live harvested products.

retailer. In the channels of distribution, the level between wholesaler and final consumer; the traditional flower shop or one who sells flowers to consumers.

rhythm. A visual flow or movement characterized by the repetition of elements or features; when repeated, color, line, form, spacing, and other elements assist the eye in moving easily throughout a design. See *repetition.*

right triangle. A form of arrangement; an asymmetrical triangle with the vertical line being perpendicular to the horizontal line, forming a 90-degree angle.

rikka (REE-kah). Literal translation: standing flowers. One of the oldest forms of Japanese flower arrangement, introduced about 1470. A large, complex form of arrangement with a kaleidoscopic view of a landscape, often used to adorn temple altars.

Roman period. During this time (28 B.C. to A.D. 325), flowers were used to make garlands and wreaths. The use of plant material was more elaborate than in the previous Greek and Egyptian periods. Fragrance and bright colors were important for flowers.

Romantic era. Name for the Victorian era in America, from about 1845 to 1900.

rose strippers. Handheld tool used for removing thorns and lower leaves from flowers, especially roses; also called dethorners.

round mound. Also called roundy-moundies, referring to circular, rounded mass bouquets. A common everyday design.

saddle. A container available in different styles and sizes, made to fit the curvature of a casket lid to hold the foundation of the casket spray design.

salesmanship. The art of selling.

sanitation. Cleanliness. The practice of keeping flower handling facilities, tools, buckets, and equipment clean.

satin acetate. Waterproof utility ribbon, available in a wide assortment of colors and widths.

satin forms. Premade satin shapes ready for floral accents to be added. Available in a variety of popular forms, such as pillows, hearts, and crosses.

saturation. See *chroma.*

scale. Another word for size. The ratio or proportion of an arrangement to the surrounding area in which it will be placed.

scalene triangle. An asymmetrical triangle in which all three sides and angles are unequal. These designs usually have a prominent vertical line, a wide angle, and a downward, diagonal line that is opposite to the height of the arrangement.

scapose. A solitary flower on a leafless peduncle or scape, such as a cut tulip or daffodil, or other inflorescence that does not have ordinary green, leafy foliage on the stem; cut gerbera, agapanthus, and anthurium are examples.

scientific name. See *botanical name, genus, Latin name* and *species.*

script. Gold lettering that is attached to a ribbon; often used for funeral designs to denote the sender's relationship to the deceased, such as "Loving Grandmother" or "Beloved Father."

sealer. An antitranspirant for fresh flowers (keeps flowers fresher longer) or a finishing spray or dip for dried flowers (keeps flowers intact).

seasonal availability. A general term referring to the peak supply times for flowers and foliage due to the plant's natural growth and flowering cycles.

seasonal flowers. Flowers that are unique to, or particularly abundant during, one season or another, such as tulips, zinnias, or poinsettias. Acacia, flowering branches, and holly with red berries are also referred to as seasonal because their availability is limited to the plants' natural growth and flowering cycles since they only grow outdoors.

secondary colors. Orange, violet, and green. Formed by mixing two primary colors together.

secondary line/stem. Term used to designate *man* line in the Ohara school.

seika (SAY-kah). Name given by some schools to the classical forms of Japanese arrangement. (Same as *shoka* in the Ikenobo school.)

sending florist. A florist who takes an order and transmits it to another florist in another city.

senescence. The aging process resulting in wilting and death of a cut flower, foliage, or plant.

sepal. A floral leaf or individual segment of the calyx of a flower; usually green. The outermost flower structure that usually encloses the other flower parts in the bud.

sequencing. A design technique in which flowers and materials are placed in a gradual or progressive change through the gradation of color, the transition of size, and so on.

sessile. Attached directly to the stem; referring to a leaf lacking a petiole or to a flower lacking a pedicel.

set piece. A funeral design that has a certain shape, usually made on a foam form, decorated with flowers, and placed on an easel. Examples include hearts, crosses, and organizational emblems.

shade. Any hue to which black has been added. A dark-colored hue.

shadowing. A design technique in which like materials are placed in pairs, one directly beneath or behind the other, to form the appearance of a shadow. Increases the feeling of depth.

shattering. The falling off of florets, leaves, or petals, especially with chrysanthemums, suffering from mechanical damage.

sheaf. A cluster of cut flowers, wheat, foliage, and the like, tied together at one end with a ribbon bow, raffia, or other tying material.

sheltered design. A design style or technique in which the floral arrangement is "protected" and contained within the container, or branches and other materials are placed over the arrangement, giving a feeling of protection and recluse.

sheltering. See *sheltered design.*

shin (sheen). Applied by most schools to the main branch, commonly called *heaven,* in a Japanese flower arrangement. Also used to designate the formal arrangement of several classical methods.

shipper. A firm that specializes in the assembly of flowers, packing in boxes, and shipping them to other buyers.

shipping. The process of transporting floral products from one point to another point, especially in the distribution channels; the speed and condition in which flowers and foliage are transported affect their quality and longevity.

shoka (SHO-kah). The latest of orthodox styles of classical arrangements. The three-branch asymmetrical, classical form of flower arrangement as practiced by the Ikenobo school. It replaced *rikka* in popularity in the late 18th century.

silica gel. A drying compound (desiccant) for use in drying flowers.

silk flowers. A general term for artificial or man-made flowers and foliage; made with realistic-looking fabrics and formed into natural shapes, and often colors, textures, and sizes. See also *everlasting flowers* and *botanicals.*

silver anniversary. The 25th anniversary; floral arrangements for this special occasion are often made in a silver-colored container with silver ribbon and a number 25 insignia accessory.

silver conditioning. See *silver thiosulfate.*

silver thiosulfate (STS). A silver compound readily taken up into the flower. Conditioning of ethylene-sensitive crops with this solution helps protect them against the effects of ethylene during handling and shipping, and increases vase life. Because of environmental concerns about silver contamination and proper disposal, its use is no longer legal in the United States. See *EthylBloc.*

skeleton flowers/foliage. The primary flowers or foliage used in setting up the framework of a design, establishing the pattern or outline of the design.

sleepiness. An ethylene-induced disorder in flowers, characterized by an inward bending, closing, or wilting of the petals. Flowers appear wilted and limp.

sleeving. A process in which flower bunches are packaged in protective waxed or unwaxed paper or polyethylene coverings to protect and separate flower heads, prevent tangling, and to identify the grower, shipper, and flower and variety name.

Society of American Florists (SAF). A national trade organization representing the needs and interests of the entire floral industry.

soe (SO-eh). Literal translation: harmonizer. Applied to secondary branch, commonly called *man,* in most methods of Japanese flower arrangement. The supporting branch in *rikka* arrangements.

Sogetsu (so-GHET-soo). Modern school of Japanese flower arrangement founded in 1926.

solitary flower. A single flower on an upright stalk; also called a terminal or axillary flower.

space. An element of design. The three-dimensional area in and around a design.

spacing. A design technique in which flowers or other materials are placed closely together to create an area of emphasis, in contrast to materials that have more negative space between them.

spadix. A spike of flowers on a swollen, fleshy axis, usually surrounded by a colorful bract, as with an anthurium.

spathe. A large bract subtended and often surrounding a spadix inflorescence, such as with a calla lily.

specialized designer. A floral designer skilled and efficient who focuses on one or several types of designs, such as casket sprays, bridal bouquets, high-style party designs, or custom silk arrangements.

specialty bouquet. A novelty bridal bouquet that is unusual or out of the ordinary, such as a muff or prayer book bouquet.

specialty flower shop. A flower shop that targets particular floral needs, such as weddings, high style designs, or everlasting designs.

species. A kind of organism. Species are designated by binominal names written in italics. See *botanical name, genus* and *Latin name.*

spike. An inflorescence in which the main axis is elongated and the flowers are sessile.

split complementary color scheme. A color scheme using one hue together with the two colors that are adjacent to the direct complement.

spool wire. Continuous wire wound on a spool, commonly used to bind large materials, create garlands, and secure foliage wreaths. See also *paddle wire.*

S-shape design. See *Hogarth curve* and *Hogarth design.*

stacking. The placing of one material on top of another.

stalk. Generally refers to a stem or peduncle.

stamen. The plant's male reproductive organ. The part of the flower producing the pollen, composed usually of anther and filament. Collectively, the stamens make up the androecium.

standard divisional pricing. A method of calculating the selling price of a floral design based on the wholesale cost of items, being a percentage of the selling price; the standard percentages and formula may vary according to a shop's financial records and the item being sold.

state florists' association. A trade association representing the needs and interests of floral industry members within a state.

static line. Lines in a composition that are parallel to the main lines of the design. For example, in a design that has predominate vertical and horizontal lines, or a rectangular format, repeating lines of vertical and horizontal display lack of visual energy and are termed static and unmoving.

stephanotis stem. A manufactured wired and taped stem that is fitted with a cotton end; the cotton end is soaked prior to fitting it into a stephanotis floret.

stigma. The region of a carpel serving as a receptive surface for pollen grains and on which they germinate.

stipule. A leafy appendage, often paired and usually at the base of the leaf stalk.

stitch wiring. A method of wiring broad leaf foliages.

stoma, *pl.* **stomata.** The pores that occur in large numbers in the epidermis of plants (stems and leaves) and through which gaseous exchanges take place.

storage cooler. An enclosed refrigeration unit in which the bulk of the flowers and foliage are stored.

strip-center flower shop. A retail floral facility that is one of several adjoining businesses that comprise a small shopping complex, usually near residential areas.

STS. See *silver thiosulfate.*

style. Slender column of tissue which arises from the top of the ovary and through which the pollen tube grows. Also, a recognizable form of design or school of thought; the end result.

stylized vertical design. A general term referring to a vertical, linear, high-style design.

subject line/stem. Designation of the *heaven* line in the Ohara school.

succulent. Plants with fleshy, water-storing stems and leaves.

symmetrical balance. Equal in visual weight on both sides of a central vertical axis.

symmetrical design. A floral arrangement that has equal visual weight on both sides of a central vertical axis; an arrangement in which the two sides are a mirror image of each other; often used for formal settings, such as in churches and funeral homes. See *symmetrical balance.*

table skirt decoration. Floral clusters or small floral accents with ribbon bows that are pinned to the skirt of the tablecloth.

tactile. Referring to the sense of touch.

tai (tie). Literal translation: material substance. Term applied by most teachers to *earth* line in classical forms and some naturalistic arrangements. The tertiary branch of the asymmetrical *shoka* style.

tailoring. Refers to trimming materials (especially leaves) to give them a sculptured, fitted, or abstract appearance.

technique. The application of design methods. A means to an end.

tension. The quality in a design created by opposing elements that offer resistance in visual flow or eye movement. For example, using discordant colors or crossing stems creates visual tension.

tepal. A perianth segment that is not clearly distinguishable as being either a sepal or petal, such as with the sepals and petals of tulips and lilies.

terracing. The placement of like materials on top of each other, but divided by space in between, giving a stair-step appearance.

terra cotta. A hard, brownish-red, usually unglazed earthenware used for containers and sculpture.

tertiary colors. Third or intermediate colors created by mixing a primary with an adjacent secondary color, named for the parent colors, such as blue-green and red-orange.

tetrad color scheme. A four-color scheme in which the four colors are equidistant from one another on the color wheel.

texture. An element of design that refers to the surface qualities or characteristics, both seen and felt, of flowers, foliage, containers, and so on.

theme. An overall feeling, style, or message that a floral design suggests, such as a floral design with a "baby girl" theme. All parts are harmonized to create the intended theme.

tint. Any color to which white has been added creating light colors and pastel hues.

tinting. A process in which cut flowers are artifically colored; flowers may be colored internally through the stem or externally by dipping or spray painting.

tipping. A general term for spray painting the edges of flower petals.

tokonoma (toh-koh-NOH-mah). Ornamental alcove in a Japanese room used to exhibit art objects and flower arrangements.

tone. Any color to which gray has been added; tones often have a gray, dusty, dull, or frosty appearance.

topiary. An evergreen tree or shrub that has been trimmed or trained to an unnatural shape. A floral design to create this appearance, such as a topiary ball or cone.

toss bouquet. A small bouquet made for the bride to throw or toss to the unmarried females after the wedding or at the reception.

total dissolved solids (TDS). Water salinity or a measure of total soluble elements in the water. A water quality characteristic, usually given in parts per million (ppm).

trade association. An organization or group that is set up by individuals, merchants, or business firms for the unified promotion and sale of flowers and floral products.

trade publication. Printed materials, such as journals, magazines, newsletters, and the like offering information of interest to all segments of the floral industry on a weekly, monthly, bimonthly, or quarterly basis.

traditional design/style. See *classical style.*

transition. A method of achieving visual rhythm. See *gradation* and *sequencing.*

transpiration. The loss of water vapor by plant parts. Most transpiration occurs through stomata.

transporter. The individual or method of shipping floral products between each level of distribution.

triadic color scheme. A color scheme using three colors that are equidistant from one another on the color wheel. The combination of red, yellow, and blue is a triadic scheme.

triangular design. A popular floral arrangement style, generally one-sided and displaying three distinct sides; may be both symmetrical and asymmetrical in form; examples include equilateral, isosceles, scalene, and right triangles.

tropism. A response (curving and bending) to an external stimulus in which the direction of the movement is usually determined by the direction from which the most intense stimulus comes. See *geotropism* and *phototropism.*

tufting. A design technique of loosely clustering or bunching materials closely together at the base of a design to emphasize color and texture. See *clustering* and *pillowing.*

tulle. A type of decorative netting used as an accessory in corsage and wedding designs.

turgid. Full of water, swollen, distended; referring to a cell that is firm due to water uptake.

tussie-mussie. A small hand-held fragrant bouquet sometimes spelled tuzzy-muzzy. The word tuzzy refers to the old English word for a knot of flowers. Originally flower cluster stems were tied together. Later, holders were manufactured for ease of carrying and displaying these tight little bouquets. During the Victorian era, flowers were symbolic and tussie-mussies conveyed sentimental messages to loved ones.

umbel. An umbrella-shaped inflorescence with all the stalks (pedicels) arising from the top of the main stem.

underwater cutter. Apparatus specially designed to cut flower stems under water, helping them to hydrate rapidly and last a longer time.

unity. A principle of design; the relationship of individual elements or parts to each other which produces a unified whole. The effect created by the cohesive placement and use of materials in which the whole is greater than its parts.

urn. A vase of classical shape with a base, often widemouthed, with two handles.

value. Measurement of the amount of light reflected from a colored object; refers to the lightness or darkness of a color.

vase life. The useful life of a cut flower after harvest, especially the lasting time and quality for the final consumer.

vegetative design. A naturalistic design style in which flowers and plant materials are placed as they would grow in nature. Materials used are found in nature together with emphasis placed on seasonal compatibility.
venation. The arrangement of veins in a leaf.
vertical design. A floral arrangement that is taller than it is wide; often used where display space is limited; examples include bud vase designs and stylized vertical designs.
vertical line. A line that has a 90-degree angle from the horizon, emphasizing strength.
vestibule. The area in a church between the outer door and the interior part or chapel of the building.
Victorian era. The period named for Queen Victoria, who reigned in England from 1837 to 1901. Floral arrangements are characterized as being massive, overdone, and flamboyant. Containers were highly decorative and gaudy.
vignette. Refers to displaying or grouping similar types of merchandise for maximum visual appeal.
visual balance. See *balance.*
visual merchandising. Efforts to attract customers to the shop and create interest in the flowers and merchandise so customers will want to buy.
void. A connecting space within a design.

wall pockets. Wall containers of various shapes for flowers.
wall spray. A composite spray made from the grouping of flat sprays and placed against the wall at funeral services.
warm colors. Colors (hues) composed of yellow, orange, and red hues. Associated with warm things like the sun and heat. Also referred to as advancing colors.
waterfall design. A design style that has many cascading layers forming a pendulous and flowing appearance.
water quality. The characteristics of water that influence its reactivity with chemicals, particularly floral preservative, and its effect on cut flowers and foliage.
water tubes. Small plastic or glass receptacles for holding water to keep flowers fresh; available in pointed or rounded forms with plastic or rubber lids that keep the water in and permit the insertion of a flower stem.
wedding consultant. A knowledgeable and experienced designer who specializes in planning wedding flowers for the bride.
Wedgwood. A fine ceramic ware popular during the English-Georgian period, named after the English potter Josiah Wedgwood. It depicted ancient Greek and Roman designs and was manufactured with special holes and openings for stems, specifically to hold flowers.
western line design. A triangular, L-shaped design with a focal point near the base from which all stems radiate. A prominent vertical line with an opposite downward sweeping line, with open space between the lines.
wetting agent. A chemical added to some floral preservatives to make water "wetter," helping to hydrate flowers quickly.
wholesaler florist/wholesaler. The traditional middleman in the channel of distribution who buys flowers and floral products from growers and brokers and sells them to retailers. Key functions are to break bulk quantities and sell smaller lots or bunches to retailers.
wilt sensitive. Flowers and foliage that are particularly sensitive or prone to water loss and wilt easily.
window display. The front window area of the floral shop that is intended to captivate the attention of people who are walking or driving by; the window area must display a powerful message or showcase merchandise so as to entice people to come into the shop.
wire. Metal that is very thin and long, like thread, with a circular cross-section; used to support fresh flower stems, replace stems (for corsage-making), bind materials, and so on; sold most commonly in 18-inch-long pieces and is available in silver and green. See also *paddle wire, spool wire,* and *chicken wire;* assigned a gauge number according to its diameter or thickness. See also *gauge.*
wire order/service. Also called "flowers-by-wire" referring to the transfer of floral design orders from one shop for delivery by another shop; originally in the early days of the floral industry, flower orders were transmitted by wire or telegraph; today customer orders are taken by one florist and sent to another by telephone, computer, or Fax.
wooden pick. A wooden stick with one end tapered to a point and having a pliable wire attached to the other end; used to secure items into a floral design or to lengthen stems.
woody stem. A tough or hard stem as opposed to a herbaceous stem; examples include heather, leptospermum, and flowering branches.
working capital. The part of a company's capital readily convertible into cash and available for paying bills, wages, and so on. The money needed to run a business.
worldwide market. The supplying, selling, and buying of floral products throughout the world due to the improvements and advancements in worldwide communication, transportation, and production.
wrap-around wiring. See *clutch wiring.*
wrapping. A design technique in which fabric, ribbon, raffia, or other materials are used to cover, coil, or twine a single stem or group of materials to achieve a decorative effect.
wreath bouquet. A specialty bouquet made in the form of a wreath carried by brides and bridal attendants.
wristlet. A device added to corsages for wearing on the wrist.

xylem. A complex vascular tissue through which most of the water and minerals of a plant are conducted.

zoning. A technique of restricting the numbers and types of materials used in certain larger areas.

Bibliography

Floral Arrangement History

Bayer, Patricia. *Art Deco Source Book.* Oxford: Phaidon Press Limited, 1988.
Benz, M. *Flowers: Free Form—Interpretive Design.* Houston, TX: San Jacinto Publishing Co., 1960.
Berrall, Julia S. *A History of Flower Arrangement.* New York: The Viking Press, 1968.
Coats, Peter. *Flowers in History.* New York: The Viking Press, 1970.
Gowing, Lawrence. *Hogarth.* London: Blading & Mansell, 1971.
Florists' Review Magazine. *A Centennial History of the American Florist.* Topeka, KS: Florists' Review Enterprises, Inc., 1997.
Maas, Jeremy. *Victorian Painters.* New York: G. P. Putnam's Sons, 1969.
Marcus, Margaret Fairbanks. *Period Flower Arrangement.* New York: M. Barrows & Company, Inc., 1952.
Mitchell, Peter. *European Flower Painters.* Schiedam, The Netherlands: Interbook International B. V., 1973.
Newdick, Jane. *Period Flowers.* New York: Crown Publishers, Inc., 1991.
Paulson, Ronald. *Hogarth: His Life, Art, and Times.* Hartford, CT: Connecticut Printers, Inc., 1971.
Schmutzler, Robert. *Art Nouveau.* New York: Harry N. Abrams, Inc., 1962.
Warren, Geoffrey. *All Colour Book of Art Nouveau.* London: Octopus Books Limited, 1972.

Floral Arrangement and Design

Allen, Judy, C. Edwards, I. Finlayson, A. Johnson, and R. Lamont. *Flower Arranging.* London: Octopus Books Limited, 1979.
Allen, Phyllis Sloan, and Miriam F. Stimpson. *Beginnings of Interior Environment.* New York: MacMillan Publishing Company, 1990.
Benz, M. *Flowers: Geometric Form.* College Station, TX: San Jacinto Publishing Co., 1986.
Better Homes & Gardens Flower Arranging. New York: Meredith Press, 1965.
Blacklock, Judith. *Flower Arranging.* Teach Yourself Books, Lincolnwood, IL: NTC Publishing Group, 1975.
Bridges, Derek. *Derek Bridges' Flower Arranger's Bible.* London: Century Hutchinson Ltd., 1990.
Brinton, Diana. *The Complete Guide to Flower Arranging.* London: Merehurst Limited, 1990.

Conder, Susan, Sue Phillips, and Pamela Westland. *The Complete Flower Arranging Book.* London: Octopus Illustrated Publishing, 1992.

Contemporary Floral Designs. Boston, MA: The Rittners Floral School, 1983.

Decorating with Plants and Flowers. Des Moines, IA: Creative Home Library, Meredith Corporation, 1972.

Dreams Come True. Lansing, MI: The John Henry Company, 1991.

Fitch, Charles Marden. *Fresh Flowers Identifying, Selecting, and Arranging.* New York: Abbeville Press, Inc., 1992.

Floral Design Techniques. Lansing, MI: The John Henry Company, 1987.

Forsell, Mary. *The Book of Flower Arranging.* New York: Michael Friedman Publishing Group, Inc., 1988.

Gatrell, Anthony. *Dictionary of Floristry and Flower Arranging.* London: The Bath Press, B. T. Batsford Limited, 1988.

Hillier, Florence Bell. *Basic Guide to Flower Arranging.* New York: McGraw-Hill Book Company, 1974.

Hillier, Malcolm. *The Book of Fresh Flowers: A Complete Guide to Selecting & Arranging.* New York: Simon and Schuster, 1988.

Hillier. *Flower Arranging.* Reader's Digest Home Handbooks. Pleasantville, NY: The Reader's Digest Association, Inc., 1990.

Inspirations of Love. Lansing, MI: The John Henry Company, 1995.

Lauer, David A. *Design Basics.* Fort Worth, TX: Holt, Rinehart and Winston, Inc., 1990.

Love, Daphne, and Sid Love. *Flower Arranging from the Garden.* London: Cassell Educational Limited, 1989.

Mann, Pauline. *The Flower Arranger's Workbook.* London: B. T. Batsford Ltd., 1989.

McDaniel, Gary L. *Floral Design and Arrangement.* Englewood Cliffs, NJ: Prentice-Hall, Inc., 1989.

Mitchell, Herbert E. *Design with Flowers.* Woodland Hills, CA: CRB Publishing, Inc., 1991.

Newdick, Jane. *Book of Flowers.* New York: Prentice Hall, 1989.

Ortho Books. *Arranging Cut Flowers.* San Ramon, CA: Chevron Chemical Company, 1985.

Parkin, Beverley. *Say It with Flowers.* Oxford: Lion Publishing, ISIS Large Print Books, 1983.

Piercy, Harold. *The Constance Spry Handbook of Floristry.* London: Christopher Helm Ltd., 1984.

The Professional Floral Design Manual. The AFS Education Center, American Floral Serivces, Inc. Okalahoma City, OK: Times-Journal Publishing Company, 1990.

Redbook Florist Services. *Advanced Floral Design.* Leachville, AR: Printers and Publishers, Inc., 1992.

Redbook Florist Services. *Basic Floral Design.* Leachville, AR: Printers and Publishers, Inc., 1991.

Redbook Florist Services. *Selling and Designing Wedding Flowers.* Leachville, AR: Printers and Publishers, Inc., 1991.

Redbook Florist Services. *Selling and Designing Sympathy Flowers.* Leachville, AR: Printers and Publishers, Inc., 1992.

Rulloda, Phillip M. *Tropical & Contemporary Floral Design.* Phoenix, AZ: Phil & Silverio, Inc., 1990.

Ryan, Tamaris. *At Home with Flowers.* New York: Gallery Books, 1990.

Sincere Sympathy. Lansing, MI: The John Henry Company, 1988.

The Tribute Collection. Lansing, MI: The John Henry Company, 1999.

Webb, Iris. *The Complete Guide to Flower & Foliage Arrangement.* Garden City, NY: Doubleday & Company, Inc., 1979.

The Wedding Flower Collection. Lansing, MI: The John Henry Company, 1997.

Holiday and Special Occasion

Ainsworth, Catherine Harris. *American Calendar Customs, Volume I.* Buffalo, NY: The Clyde Press, 1979.

Ainsworth, Catherine Harris. *American Calendar Customs, Volume II.* Buffalo, NY: The Clyde Press, 1980.

Amer, Jean B. *Flower Arrangements for Special Occasions.* Nashville, TN: Allied Publications, Inc., 1962.

Chase's Annual Events. Chicago: Contemporary Books, Inc., 1989.

Douglas, George William. *The American Book of Days.* New York: The H. W. Wilson Company, 1948.

Fendelman, Helaine, and Jeri Schwartz. *Official Price Guide Holiday Collectibles.* New York: House of Collectibles, 1991.

Redbook Florist Services. *Floral Design for the Holidays.* Leachville, AR: Printers and Publishers, Inc., 1991.

Floral Fragrance

Bonar, Ann. *Gardening for Fragrance.* London: Ward Lock Limited, 1990.

Duff, Gail. *Natural Fragrances.* Pownal, VT: Storey Communications, Inc., 1989.

Lacey, Stephen. *Scent in Your Garden.* Boston: Little, Brown & Company Limited, 1991.

Ohrbach, Barbara Milo. *The Scented Room.* New York: Clarkson N. Potter, Inc., 1986.

Taylor, Jane. *Fragrant Gardens.* London: Ward Lock Ltd., 1991.

Nomenclature and Postharvest Physiology

Bailey, Liberty Hyde, and Ethel Zoe Bailey. *Hortus Third.* New York: MacMillan Publishing Co., Inc. 1976.

Benson, Lyman. *Plant Classification.* Boston: D. C. Heath and Company, 1957.

Capon, Brian. *Botany for Gardeners.* Portland, OR: Timber Press, 1990.

Coombes, Allen J. *Dictionary of Plant Names.* Portland, OR: Timber Press, 1987.

Heywood, V. H. *Flowering Plants of the World.* New York: Mayflower Books, Inc., 1978.

Kays, Stanley J. *Postharvest Physiology of Perishable Plant Products.* New York: Van Nostrand Reinhold, 1991.

Lawrence, George H. M. *An Introducton to Plant Taxonomy.* New York: The MacMillan Company, 1956.

Porter, C. L. *Taxonomy of Flower Plants.* San Francisco: W. H. Freeman and Company, 1967.

Radford, Albert E. *Fundamentals of Plant Systematics.* New York: Harper & Row, Publishers, Inc., 1986.

Raven, Peter H., and Ray F. Evert. *Biology of Plants, 3rd Edition.* New York: Worth Publishers, Inc., 1981.

Salisbury, Frank B., and Cleon W. Ross. *Plant Physiology, 3rd Edition.* Belmont, CA: Wadsworth Publishing Company, 1985.

Seymur, Edward L. D. *The Wise Garden Encyclopedia.* New York: Harper Collins Publishers, 1990.

Care and Handling

Graber, Debra Terry. *Fresh Flowers Book 2.* Lansing, MI: The John Henry Company, 1989.

Holstead-Klink, Christy. *Care and Handling of Flowers and Plants.* Alexandria, VA: Society of American Florists, 1985.

McKinley, William J. *The Cut Flower Companion.* Danville, IL: Interstate Publishers, 1994.

Redbook Florist Services. *Purchasing and Handling Fresh Flowers and Foliage.* Leachville, AR: Printers and Publishers, Inc., 1991.

Vaughan, Mary Jane. *The Complete Book of Cut Flower Care.* London: Christopher Helm Ltd., 1988.

Everlasting Flowers

Conder, Susan. *Dried Flowers.* Boston: David R. Godine, Publisher, Inc., 1988.

Condon, Geneal. *The Complete Book of Flower Preservation.* Englewood Cliffs, NJ: Prentice-Hall, Inc., 1972.

Foster, Maureen. *The Flower Arranger's Encyclopedia of Preserving and Drying.* London: Blandford, 1988.

Hillier, Malcolm, and Colin Hilton. *The Book of Dried Flowers.* New York: Simon and Schuster, 1986.

Penzner, Diana, and Mary Forsell. *Everlasting Design.* Boston: Houghton Mifflin Company, 1988.

Petelin, Carol. *The Creative Guide to Dried Flowers.* London: Webb & Bower Limited, 1990.

Oriental Style of Design

Berrall, Julia S. *A History of Flower Arrangement.* New York: The Viking Press, 1968.

Conway, J. Gregory. *Flowers: East–West.* New York: Alfred A. Knopf, Inc., 1938.

Davidson, Georgie. *Classical Ikebana.* New York: A. S. Barnes & Company, 1970.

Kawase, Toshiro. *Inspired Flower Arrangements.* New York: Kodansha International Ltd., 1990.

Marcus, Margaret Fairbanks. *Period Flower Arrangement.* New York: M. Barrows & Company, Inc., 1952.

Sparnon, Norman J. *Japanese Flower Arrangement.* Tokyo: Charles E. Tuttle Company, 1960.

Webb, Lida. *An Easy Guide to Japanese Flower Arrangement Styles.* New York: Hearthside Press, Inc., 1963.

Wood, Mary Cokely. *Flower Arrangement Art of Japan.* Tokyo: Charles E. Tuttle Company, 1959.

Harvest and Distribution

Graber, Debra Terry. *Fresh Flowers Book 2.* Lansing, MI: The John Henry Company, 1989.

Guide to Floral Industry Transportation. Alexandria, VA: Society of American Florists, 1988.

Redbook Florist Services. *Purchasing and Handling Fresh Flowers and Foliage.* Leachville, AR: Printers and Publishers, Inc., 1991.

U.S. Department of Agriculture website: http://www.usda.gov/ nass

The Retail Flower Shop

Cavin, Bram. *How to Run a Successful Florist & Plant Store.* New York: David McKay Company, Inc., 1977.

McDaniel, Gary L. *Ornamental Horticulture.* Reston, VA: Reston Publishing Co., 1979.

Pfahl, Peter B., and P. Blair Pfahl, Jr. *The Retail Florist Business.* Danville, IL: The Interstate Printers & Publishers, Inc., 1983.

Redbook Florist Services. *Retail Flower Shop Operation.* Leachville, AR: Printers and Publishers, Inc., 1991.

Royer, Kenneth R. *Retailing Flowers Profitably.* Lebanon, PA: Royer Publishing, 1998.

Society of American Florists website: http://www.safnow.org

Sullivan, Glenn H., Jerry L. Robertson, and George L. Staby. *Management for Retail Florists.* New York: W. H. Freeman and Company, 1980.

Cut Flowers and Foliage

Bailey, Liberty Hyde, and Ethel Zoe Bailey. *Hortus Third.* New York: MacMillan Publishing Co., Inc., 1976.

Bianchini, Francesco, and Azzurra Carrasa Pantano. *Guide to Plants and Flowers.* New York: Simon & Schuster Inc., 1974.

Bloemen Bureau Holland. Leiden, The Netherlands: Flower Council of Holland, 1992.

Cut Flower Guide. Gold River, CA: California Cut Flower Commission, 1997.

Fitch, Charles Marden. *Fresh Flowers Identifying, Selecting, and Arranging.* New York: Abbeville Press, Inc., 1992.

Graber, Debra Terry. *Fresh Flowers Book 2.* Lansing, MI: The John Henry Company, 1989.

Halpin, Anne. *The Naming of Flowers.* New York: Harper & Row, Publishers, Inc., 1990.

Hay, Roy, and Patrick M. Synge. *The Color Dictionary of Flowers and Plants for Home and Garden.* New York: Crown Publishers, Inc., 1991.

Heywood, V. H. *Flowering Plants of the World.* New York: Mayflower Books, Inc., 1978.

Hodgson, Margaret, Roland Paine, and Neville Anderson. *A Guide to Orchids of the World.* Prymble, Australia: Collins Angus & Roberston Publishers, 1991.

Holstead-Klink. *Care and Handling of Flowers and Plants.* Alexandria, VA: Society of American Florists, 1985.

Larson, Roy A. *Introduction to Floriculture.* San Diego, CA: Academic Press, Inc., 1992.

Laurie, Alex, D. C. Kiplinger, and Kennard S. Nelson. *Commercial Flower Forcing.* New York: McGraw-Hill Book Company, 1979.

McKinley, William J. *The Cut Flower Companion.* Danville, IL: Interstate Publishers, Inc., 1994.

New Cut Flower Crops. Grower Guide, No. 18. London: Grower Books, 1982.

New Pronouncing Dictionary of Plant Names. Chicago: Florists' Publishing Company, 1990.

Raven, Peter H., and Ray R. Evert. *Biology of Plants, 3rd Edition.* New York: Worth Publishers, Inc., 1981.

Reader's Digest Encylcopaedia of Garden Plants and Flowers. London: The Reader's Digest Association Limited, 1975.

Seiden, Allan. *Flowers of Aloha.* Aiea, HI: Island Heritage Publishing, 1990.

Seymour, Edward L. D. *The Wise Garden Encyclopedia.* New York: Harper-Collins Publishers, 1990.

Vaughan, Mary Jane. *The Complete Book of Flower Care.* London: Christopher Helm Ltd., 1988.

Periodicals

Dateline: Washington. Alexandria, VA: Society of American Florists.

Design with Flowers. Costa Mesa, CA: Herb Mitchell Associates, Inc.

Floral Finance. Tulsa, OK: American Floral Services, Inc.

Floral Management. Alexandria, VA: Society of American Florists.

Floral Mass Marketing. Chicago, IL: Cenflo, Inc.

Florist Magazine. Livonia, MI: Florists' Transworld Delivery Association.

Florists' Review. Topeka, KS: Florists' Review Enterprises, Inc.

Flower News. Chicago, IL: Cenflo, Inc.

Flowers &. Los Angeles, CA: Teleflora.

Holland Flowers. New York: The Flower Council of Holland.

PFD (Professional Floral Designer). Oklahoma City, OK: American Floral Services, Inc.

The Retail Florist. Oklahoma City, OK: American Floral Services, Inc.

Supermarket Floral. Overland Park, KS: Vance Publishing.